Praise for *An End to Upside Down Thinking*

I love this book. An End to Upside Down Thinking *will get you thinking. This book is full of positive, insightful and powerful information.*

—**Jack Canfield**, *New York Times* bestselling author of *The Success Principles* and co-creator of *Chicken Soup for the Soul* book series

The scientific study of the mind-brain relationship, including all manner of human experiences, combined with the deepest mysteries of modern physics, are leading to an unprecedented shift in human understanding of the nature of reality, one that many in the field feel will make the Copernican Revolution seem minuscule by comparison. In An End to Upside Down Thinking, *Mark Gober provides a broad sketch of the relevant scientific lines of inquiry to support this inevitable, yet very empowering and optimistic, shift in understanding of the nature of human existence. Especially as one realizes the damage that has been done by our reigning materialistic paradigm and its false sense of separation, the promise of this newest unifying scientific revolution becomes clear—we must grow into this new understanding, if for no other reason than to survive.*

—**Eben Alexander, MD**, neurosurgeon and author of *Living in a Mindful Universe* and the #1 *New York Times* bestseller *Proof of Heaven*

Mark Gober's question sums up the entire conundrum about what we thought we knew but turned out to be false or at least highly questionable knowledge: If consciousness isn't just a product of the brain, and if it survives the death of the physical body—beyond space and time—then how does it fit into our picture of the universe? We thought we knew what the real world is really like, namely that it is material and that consciousness is the product of a material brain. But what if the world isn't, and the brain isn't? What kind of a world is it then, and what kind of consciousness do we have then? Mark seeks an answer in clear and meaningful terms. A great service to the cause of advancing knowledge and its proper application to our thinking and our life. A big vote of thanks to him for helping to bury the outdated, but not yet outmoded, materialism of our time.

—**Dr. Ervin Laszlo**, two-time Nobel Peace Prize nominee, author of more than 75 books, philosopher of science and systems theorist

Ever wonder why you are thinking of someone and moments later they call or connect to you? Magic? Coincidence? Maybe not. New information on what we call phenomena is brilliantly shared in this must-read book! New pathways of the mind will open and perhaps shift your perception of reality, time and space and above all consciousness! Bravo.

—**Goldie Hawn**

What if our most basic assumptions about our understanding the mind are wrong, upside down? Mark Gober has created a remarkable synthesis about one of the most challenging areas to understand, the infamous 'hard problem' of consciousness. To be humble about current conceptions of how our mind works, and open minded to alternative perspectives than pure materialism, has the potential to promote a leap in scientific discovery and possibly even a more compassionate society.

—**Elissa Epel, PhD**, Professor, Department of Psychiatry, UCSF; co-author of the *New York Times* bestseller *The Telomere Effect*

Almost everything you learned in school about who and what you are is wrong. In An End to Upside Down Thinking, *Mark Gober, a Princeton-trained financier and hardcore rationalist, describes his shock at discovering that this is not some wild conspiracy theory, but an open secret supported by solid scientific evidence. If you'd like to know what was left out of your college education, there's no better place to begin than this easy-to-read survey of the mind-boggling nature of reality and your place in it.*

—**Dean Radin, PhD**, chief scientist at the Institute of Noetic Sciences

Revolution is in the air, and Mark Gober is out front carrying a banner in his An End to Upside Down Thinking. *He turns things right side up, explaining why consciousness is fundamental, nonlocal, and cannot be explained in terms of brain mechanisms. This is one of the most incisive indictments of the materialist view of consciousness I've read. Gober's message is urgent; he shows why our future depends on this course correction. Viva la revolución!*

—**Larry Dossey, MD**, author of *ONE MIND: Why our Individual Mind is Part of a Greater Consciousness and Why It Matters*

Thoughts and realities that we thought were rock solid are shaken to the core in this terrific new book. An exposé that will force the reader to look at their individual assumptions about the very nature of how events happen and will continue to happen and be shocked that the universe isn't as orderly as they have always thought. You will think about what has been written here for many many months after you have read the last chapter.

—**Barry Baker**, Senior Advisor at Lee Equity

Mark Gober's An End to Upside Down Thinking *shows with broadly extensive examples how the conventional 19th century common sense dogmatic materialistic worldview has no basis in fact. Scientific revolutions always begin with a steady accumulation of anomalous evidence. The consciousness-as-reality revolution will likely be difficult as the current paradigm is deeply entrenched in seats of authority and power. This book promises to open the minds of many potential contributors to the growing mountain of anomalies and to the highly imaginative among us that can ground these puzzling phenomena in a comprehensive and clear theory. Send a copy to everyone you know.*

—**Loren Carpenter**, computer scientist, founder of Pixar, and two-time Academy Award winner

In An End to Upside Down Thinking, *Mark Gober presents a comprehensive overview of the evidence threatening the core assumptions of the prevailing materialistic view of reality, where consciousness is regarded as a product of the physical brain. He makes a compelling argument for the need of a radical transformation of science and the need to recognize consciousness as the primary organizing principle of the Universe and the ultimate source of reality.*

—**Brenda Dunne**, President, ICRL, and former Princeton Engineering Anomalies Research Laboratory (PEAR) manager

This book changes everything. You will re-examine your entire belief system and formulate a new framework for creating meaning in your life. Mark brilliantly connects scientific theories, research, and human experience to examine consciousness. Realizing that you may be dismissing or discounting information you are accessing through your consciousness has practical implications to enhance your life. You will trust your "intuition" more fully, feel more connected to others, be happier with yourself and be inspired. As the number of readers increases, the world will be a better place! Thank you, Mark!

—**Ann Shippy, MD**, author of *Mold Toxicity Workbook* and *Shippy Paleo Essentials*

Modern "secular humanism," a quasi-religious worldview that arose in connection with the classical physics of the late 19th century, holds that subatomic physical things of some sort are the sole ultimate realities and that all else including mind and consciousness is derivative. This bleak vision currently constitutes the received wisdom of opinion elites throughout the civilized world; it dominates contemporary social sciences, psychology and neuroscience, has destructively colonized neighboring academic disciplines including most of the humanities (even religious studies!) and rules our media and educational systems from top to bottom. Despite its associated practical achievements, this worldview is now known to be false at its very foundations, and it is in the process of being overthrown in favor of a more comprehensive science-based metaphysical vision which makes consciousness the ultimate reality and is capable of accommodating our deepest spiritual experiences and yearnings. In this lively, provocative and well-written survey, Mark Gober provides a sampling of the many kinds of scientific findings that are driving this tectonic shift in worldviews, and a thoughtful discussion of its profound implications for our individual and collective human fates.

—**Edward F. Kelly, PhD**, lead author of *Irreducible Mind* and *Beyond Physicalism*, and Professor of Psychiatry and Neurobehavioral Sciences at the University of Virginia

A comprehensive overview. Well done. I hope it has a big impact.

—**Rupert Sheldrake, PhD**, former Cambridge University biochemist; author of *Science Set Free* and *Morphic Resonance*

In An End to Upside Down Thinking, *Gober retrofits science with the tools it needs to move forward in the new world of empirical results that don't make sense according to materialist ways of thinking. By bringing the reader expertly, engagingly, and accessibly toward a new and more accurate understanding of the nature of the universe, he addresses the mysteries of time, space, and causality in an updated and decidedly revolutionary way.*

—**Julia Mossbridge, MA, PhD**, cognitive neuroscientist and director of the Innovation Lab at the Institute of Noetic Sciences

Mark Gober's book is an insightful, meticulous, and amazing investigation of consciousness. He takes the NDE (near death experience) to another level. Psychic occurrences once considered strange, awkward or even taboo must be intertwined into our present daily life. We cannot just deny them any longer! As a cardiologist who has dealt with multiple situations of life/death, and life again, I can attest that this is a very important and extraordinary book. This is a must read that will assist us in the 21st Century with an amazing "re-awakening."

—**Stephen T. Sinatra, MD, FACC**; co-author, *Health Revelations from Heaven*

To listen to evidence requires an open mind. To then follow wherever it leads often requires courage, persistence and honesty. Having earlier accepted the current mainstream paradigm of materialistic science that states that mind and consciousness arise from matter, Mark Gober has undertaken a tremendous journey of inner and outer discovery, turning his previous world view upside down and revealing just how wrong mainstream science has been. In An End to Upside Down Thinking, *he brilliantly investigates numerous and compelling evidence by multiple researchers and across a broad range of phenomena to show that consciousness isn't something we somehow have but what we and the whole world are. The revolution has begun!*

—**Dr. Jude Currivan**, cosmologist; author of *The Cosmic Hologram: In-formation at the Center of Creation*

For someone to write a book such as An End to Upside Down Thinking, *on topics dared to be included in this manuscript, is both essential and courageous in this moment of extensive and planned human ignorance. Out of the halls of centuries of information suppression come these chapters—like magical doors to temples of radical wisdom. The pages of* An End to Upside Down Thinking *are antidotes to many ailments spread worldwide, where humans are worked like slaves rather than viewed as windows into the consciousness of GOD. Congratulations to Mark Gober for such an important and uplifting undertaking.*

—**Guru Singh**

I have read several books among this genre, yet this one—from the first chapter—has the reader questioning many normal assumptions. I predict this book will be widely perceived as important.

—**Gregory Miller**, Of Counsel, Wilson Sonsini Goodrich & Rosati; former managing director, Google.org

Mark Gober's masterful book, An End to Upside Down Thinking, *is poised to create worldwide impact by redefining what it means to be human. Gober ultimately raises a question that is critically important in today's turbulent global climate: if we are truly and fundamentally interconnected, how should we treat one another? The implications extend to all aspects of our culture—from science, to politics, to education, to gender equality, and beyond.*

As Founder & CEO of Mogul, which enables millions of women worldwide to connect, share information, and access knowledge from each other across 196 countries and 40,000 cities, this book provides a new scientific lens through which equality and empowerment are only natural since we are all interconnected and part of the same whole. The time is now for a collective mindset shift at this pivotal juncture in human history; Gober's paradigm-shifting book is a catalyst that society desperately needs.

—**Tiffany Pham**, founder and CEO of Mogul

Humanity is at a critical turning-point moment in our history. In every area of society, old paradigms are crumbling. While the materialistic/mechanistic world-view gave us many gifts, we are also rapidly destroying our planet. Technology has been advancing, yet people are chronically stressed out, sick, unfulfilled, and unhappy. It's time for change. Fortunately, there's a new model of reality being birthed by pioneering scientists all around the world, one that honors the interconnectedness of all things, mind-matter interactions, the power of consciousness to influence physical reality, and the holistic nature of our human experience. The implications are massive across all sectors of society. A revolution is underway! Mark Gober's An End to Upside Down Thinking *is the single most succinct and comprehensive tour of this exciting new paradigm. In my 20 years of exploring consciousness, metaphysics, science and spirituality, I've never read a book that's so thoroughly-researched and so daring in its scope. Travel into the frontiers of science. Experience the new realm of human possibilities. Read this book!*

—**Dr. Edith Ubuntu Chan**, holistic medicine doctor and human potential expert, #1 bestselling author of *SuperWellness: Become Your Own Best Healer*: www.DrEdithUbuntu.com/www.SuperWellness.com

A riveting exploration of existing scientific evidence for paranormal realities. An important reminder of how much of our world is still left to be explained, and how powerful our human potential truly is.

—**Giancarlo Marcaccini**, CEO of Yogi Tea

Mark Gober is clear about his intention for this book. He wants to make some very interesting yet unheralded scientific findings available to a wide audience of readers. The science is good, but the topic is taboo, though it seems that status may be slowly changing. We're talking about consciousness, and more particularly, its status in a scientific framework that struggles to find a viable definition of that most intimate and directly experienced aspect of the world. Gober lays out the questions and the tentative answers, and goes for the science that has some chance of nailing down what has to be understood for us to come to grips with mind and its extraordinary presence in the world. Here you will find a wide-ranging picture, painted in primary colors, of psi research, studies of telepathy, psychokinesis, precognition, and near-death experiences, studies that may ultimately make the difference in understanding what consciousness is and what it does—while providing direct access to our world. In terms of a quotation from the book: "For these [phenomena] we need a new language altogether, as we need new theories from a new kind of science even to begin to comprehend them." With that new language we can speak to each other of the hugely important understanding that we are not separate islands of individuality, but participants in a connected whole.

—**Roger Nelson, PhD**, Director of the Global Consciousness Project and former Coordinator of Research at the Princeton Engineering Anomalies Research Lab (PEAR)

An End to Upside Down Thinking

Dispelling the Myth That the Brain Produces Consciousness, and the Implications for Everyday Life

Mark Gober

Printed in the United States of America
First Printing, 2018

ISBN-13: 978-1-947637-85-6 print edition
ISBN-13: 978-1-947637-86-3 ebook edition
ISBN-13: 978-1-947637-87-0 audio edition

Waterside Press
2055 Oxford Ave.
Cardiff-by-the-Sea, CA 92007
www.waterside.com

To the individuals who have bravely departed from orthodoxy to explore the limits of science. Without their efforts, this book would not have been possible.

All I did was collect and organize the pieces.

"Everything we know is only some kind of approximation," Richard Feynman once said. "Therefore, things must be learned only to be unlearned again or, more likely, to be corrected." This is where Galileo, Newton, Darwin, and Einstein did their work. All the revolutionaries have been challenged, accepted, then challenged again. As George Bernard Shaw put it, "All great truths begin as blasphemies."

Where science does have a problem is in the fact that our collective memories are so short. Once that resigned acceptance of a discovery comes, we forget that there was once such a kerfuffle. We act as if truth were always with us, that it is self-evident. We forget the decades of persecution someone endured in order to shepherd us to the view we would now die to defend. And so we become comfortable—so comfortable that we will wantonly persecute the man or woman who comes to disturb our newfound peaceful state.

—Quantum physicist Michael Brooks,
author of "Beyond the Safe Zones of Science"
(*EdgeScience*, September 2015)

Contents

Preface

A framework to consider while reading this book

Before you begin reading, I warn you that you might need to suspend everything you thought you knew about reality. Remind yourself that while humanity has come a long way, there is still *a lot* that we do not know. For example, ~96 percent of the universe is mysterious "dark matter" and "dark energy," about which we know very little. As billionaire hedge-fund manager Ray Dalio advises in his *Principles*: "I believe you must be radically open-minded."[1] I likewise encourage radical open-mindedness as we contemplate theories of our existence in this book.

Whether you realize it or not, most of modern society's thinking is based upon a philosophy known as "materialism"—the notion that physical material, known as "matter," is fundamental in the universe. In other words, matter is the basis of all reality. Everything is comprised of matter, and everything can be reduced to matter.

The basic thinking is as follows: There was a "Big Bang" 13.8 billion years ago that started the universe. Units of matter—atoms—interacted throughout the universe. The interactions of matter are commonly called "chemistry." After countless, random chemical reactions, self-replicating molecules known as DNA eventually formed on Earth. DNA molecules served as the building blocks for the evolution of life. Human beings and other organisms evolved and developed brains. The brain enabled humans to have minds and awareness: an "inner experience" sometimes called "consciousness."

In short, materialism assumes that matter (e.g., the brain) produces consciousness, as shown in Figures A and B on the following pages.[2]

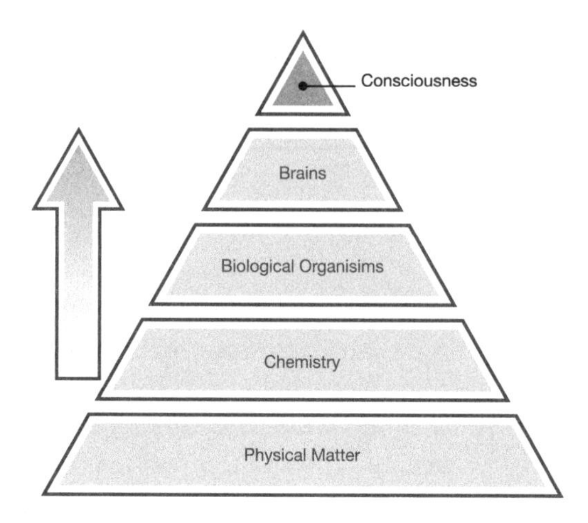

Figure A. Today's mainstream scientific view, known as materialism, is that consciousness is produced by the brain, which is a product of physical matter

This line of thinking informs views on our existence. Because materialism assumes that the brain produces consciousness, when your brain dies, your consciousness dies. If there is no brain, then there is no consciousness. Therefore, any meaning a person ascribed to his or her life while living is wiped out once the person dies.

That might sound bleak and nihilistic, but unfortunately it is what a literal interpretation of materialism implies. I know this because I used to be a materialist; and as someone who relied strictly on logical reasoning rather than faith, I had no choice but to accept these implications.

However, simple introspection revealed that materialism is a superstitious belief system that cannot be proven.

Here's why.

Think of any experience you have. "I am happy." "I am sad." "I see the car." "I feel a burning sensation." Constant in those experiences—in *any* experience—is "I": the subject that is experiencing something. It is not possible to definitively verify an experience without an "I" (i.e., consciousness) to experience it.

Imagine a universe in which all conscious beings were absent. Could that universe exist? It is possible. Materialism would predict that the world would go on merrily without any form of consciousness. But can we prove that? Technically, no, we cannot *prove* it. If there were no conscious observers, there would be no living beings to *confirm* that anything existed.[3]

Therefore, materialism, which assumes that matter comes before consciousness, is an unverifiable belief system. How can it be *proven* that anything exists or existed without some "I" to experience it? Without consciousness, we cannot prove that anything exists.

As philosopher Rupert Spira puts it: "The materialist perspective is not grounded in experience. It requires an abstract line of reasoning that presupposes the existence of a reality outside consciousness, although nobody has ever experienced this, nor could they ever experience it. The materialist point of view asserts the reality of that which is *never* experienced—matter [outside consciousness]—and denies that which alone is *always* experienced—consciousness itself. That is the tragedy and the absurdity of the materialist perspective from which humanity is suffering"[4] [emphasis in original].

Materialism's logic is twisted when you break it down. You might need to read this slowly.

- Matter's existence before consciousness, as described above, cannot be known for sure. We just showed that. Therefore, matter's existence before consciousness is "unknown." In other words, it is an abstraction.
- Conversely, we know that we have conscious experience—you are conscious as you read these words. So, consciousness is unquestionably "known." In other words, it is concrete.

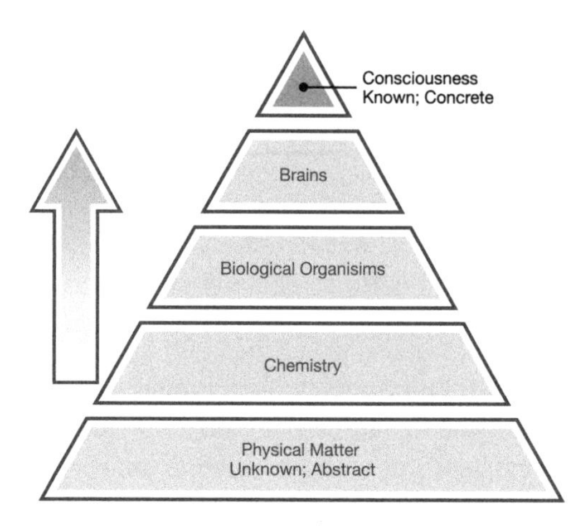

Figure B. Materialism purports that matter's existence before consciousness (an unknown abstraction) creates consciousness (which is known and concrete)

Now, let's reexamine materialism with this lens. Materialism assumes that matter precedes and creates consciousness. We just established that matter's existence before consciousness is unknown, whereas, with consciousness it is known. So materialism is saying, "Let's use an unknown, abstract thing to infer the known, concrete thing."

Most areas of logical inquiry *start with the knowns* to explore unknowns. Materialism has it backward. It says the unknown causes the known, that the abstraction causes the concrete.[5]

For these reasons, among others, philosopher Bernardo Kastrup, PhD, entitled his 2014 book *Why Materialism Is Baloney*. As he states it: "Materialism is a reasonable castle built on top of rotten foundations."[6]

Why is this important? Modern science, which is dominated by materialism, prides itself on evidence and proof. It often criticizes religions for relying on leaps of faith to justify beliefs. For example, materialist biologist Richard Dawkins ridicules faith. In his words: "What is faith but belief without evidence? Faith is...believing it because you want to believe it.... That is not a respect-worthy reason to believe anything."[7] Ironically, the basic tenet of materialism—which is that matter (the brain) produces the consciousness—is based on its own leap of faith. There is no controlled, double-blind study that science can conduct to conclusively prove that matter preexisted consciousness.

Yet materialist science seems to overlook the shaky grounds on which its foundation sits. We aren't warned of the above questions in high school science class. We aren't told about the leap of faith we are unknowingly taking. But one of the greatest scientists in history, Albert Einstein, explicitly acknowledged this issue in a conversation in 1930 with Bengali mystic and Nobel Prize winner Rabindranath Tagore. Einstein was a materialist, believing in a world independent of consciousness, but humbly admitted that his framework was not provable: "I cannot prove that my conception is right, but that is my religion."[8] That's right, folks, Einstein said it: The foundation of modern science—ironically—is a form of religion.

The irony goes even further. Many contemporary skeptics in science are also materialists. If materialism is based on a leap of faith, then shouldn't a true skeptic be skeptical of materialism? Shouldn't skeptics be skeptical of their own materialist belief system?

Not only are there logical and philosophical questions about whether a material brain produces consciousness, but there are also serious scientific questions. We don't know how physical matter gives rise to a nonphysical mind.

Let's do a quick exercise.

> Please touch your arm.
> Now, touch your leg.
> Now, touch your mind.
> You can't touch your mind.

How does a physical body that you can touch—produce a nonphysical mind—that you can't touch? As physicist Peter Russell puts it: "How can something as immaterial as consciousness ever arise from something as unconscious as matter?"[9]

This question even has a special name in science and philosophy. It is known as the "hard problem"[10] of consciousness. How hard is it? So hard that *Science* magazine's 125th-anniversary edition (2005) listed this question second among its top 25 remaining questions in all of science. It poses the question: "What is the biological basis of consciousness?"[11] We do not know the answer. The responses we get from science are often similar to neuroscientist Sam Harris's statement: "There is nothing about a brain, studied at any scale, that even *suggests* that it might harbor consciousness"[12] [emphasis in original]. Philosopher Christian de Quincey further sums up the situation: "Scientists are in the strange position of being confronted daily by the indisputable fact of their own consciousness, yet with no way of explaining it."[13]

No one would dispute that the brain is *related* to consciousness. However, we have no evidence that the brain *produces* consciousness. Let's look at an analogy to elaborate on this idea. When there is a large fire, many firefighters are often present. But we do not conclude that because there are many firefighters at the site of the fire that the firefighters caused the fire.[14] The fact that two concepts are related or co-occur does not always mean that one causes the other. Similarly, we can't automatically conclude that the brain must produce consciousness simply because the brain and consciousness are related.

As we will explore in this book, a branch of physics first conceived in the early 20th century—known as quantum mechanics—provides a picture of reality that shatters common sense and puts materialism into question. Its findings led Nobel Prize-winning physicist Max Planck to declare in 1931: "I regard consciousness as fundamental. I regard matter as derivative from consciousness. We cannot get behind consciousness. Everything we talk about, everything that we regard as existing, postulates consciousness."[15]

And as stated by another early 20th century quantum physicist, Sir James Jeans: "Mind no longer appears to be an accidental intruder into the realm of matter...we ought rather hail it as the creator and governor of the realm of matter."[16]

Planck's and Jeans's perspectives move consciousness to the base of the triangle shown previously in Figures A and B, while preserving the integrity of everything else. We do not need to give up what we have learned in physics, chemistry, biology, or neuroscience; we are simply recontextualizing these disciplines. All we do is invert the order: consciousness comes first, not matter. Matter, chemistry, biological organisms, and brains all still exist, but they exist *within* consciousness.

While the idea might sound radical, it is actually a more skeptical framework than materialism because it *starts* with the "known"—the most obvious and undeniable part of our existence: consciousness (see Figure C next page).

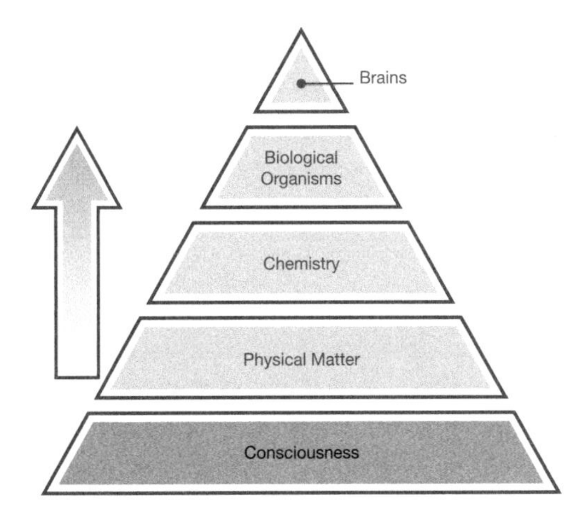

Figure C. An alternative perspective suggesting consciousness is fundamental and everything else (e.g., physical matter and the universe…and even brains) comes from, and is experienced within, consciousness

Put another way by English philosopher F. C. S. Schiller: "Materialism is…a putting of the cart before the horse, which may be rectified by just *inverting* the connexion [sic] between Matter and consciousness. Matter is not that which *produces* consciousness, but that which *limits* it and confines its intensity within certain limits"[17] [emphasis in original].

If consciousness is indeed more fundamental than matter, as Planck, Jeans, Schiller, and others suggest, then the "paranormal"—"anomalies" in science, such as psychic abilities and the mind's survival of bodily death—become expected. They are only paranormal or anomalous if we assume that consciousness comes from the brain. They make no sense in that context. But if consciousness is the foundation of reality, then it would be easy to imagine, for instance, that consciousness could flow from one person to another (telepathically). And since the body is just a product of consciousness, the body's death would not imply that its consciousness also dies.

Dr. Kastrup provides a metaphor to explain how the theory works. Imagine that all of reality is a stream of water, where water represents consciousness. Now imagine that within the stream, whirlpools form. Those whirlpools are self-localizations of water. While they may look different from other parts of the stream, the whirlpools are made of nothing more than water.

In Dr. Kastrup's analogy, whirlpools represent individuals in a stream of consciousness. My brain is one whirlpool, yours is another, etc. Since water is the fundamental medium of the stream, it is possible that sometimes water from one whirlpool can end up in another whirlpool (think: psychic abilities). And when a whirlpool dissipates, the water simply flows into the broader stream (think: consciousness continues when the physical body dies).

Under this framework, we can understand why *Science* magazine's #2 question: "What is the biological basis of consciousness?" hasn't been solved—it is the wrong question! If consciousness exists independently of the brain, then of course we wouldn't be able to find a biological basis of consciousness. The answer to *Science* magazine's question is: There is *no* biological basis of consciousness in the first place.

As stated by Dr. Kastrup: "To say that the brain *generates* mind is as absurd as to say that a whirlpool generates water!"[18] [emphasis in original]. So we might have just cracked the "hard problem" of consciousness. And incidentally, the above framework answers *Science* magazine's #1 remaining question in science: "What is the universe made of?"[19] The answer, according to the described framework: "Consciousness."

I urge you to keep these ideas in mind (no pun intended) as you read this book. We will discuss a host of phenomena that might seem impossible. The seemingly unbelievable becomes believable if we suspend the materialist religion we have been taught and remain "radically open-minded" to the notion that consciousness is more fundamental than matter.

If even *one* of the anomalous phenomena described in this book is, in fact, real—which I have become convinced is the case—then the "Consciousness is primary" framework is a much more suitable picture of reality than is materialism. And if that is true, then we need to rethink—together, as a civilization—science, technology, medicine, education, politics, and what it means to be human.

Section I
Introduction

CHAPTER 1

Introducing the Author and the Book's Contents

> *All truth goes through three stages. First it is ridiculed. Then it is violently opposed. Finally it is accepted as self-evident.*[1]
>
> —Arthur Schopenhauer, 19th-century German philosopher

What if I told you that some people know when they are being stared at *from behind*, and they even know when they are being stared at when the "looker" is staring from another room *by video camera*?

Or if I told you that some twins know when the other twin is in danger even when they are physically separated; and some twins feel physical pain when the *other* twin is injured?

Or if I told you that some dogs know when their owners mentally *decide* to come home, even when the owners come home in randomly selected taxis, from randomly selected locations (miles away), at randomly selected times?

Or if I told you that people's bodies respond to erotic images seconds *before* the image is randomly generated by a computer, even though the person is unaware that the computer will show an erotic image?

Or if I told you that the U.S. government ran a program led by laser physicists for more than 20 years in which people (called "remote viewers") "sent their minds" to distant locations and described what they saw in the past, present, and future, and successes were confirmed by CIA documents released to the public in 2017 (among other sources)?

Or if I told you that some patients who undergo cardiac arrest (i.e., their heart stops and there is no brain function), upon being brought back to life recall lucid memories and logical thought processes that occurred *during the time of no brain function*?

Or if I told you that some people who have been blind since birth who have a "near-death experience" report that during the experience they are able to see, but after the experience they go back to being blind?

Or if I told you that a savant, deemed by doctors as "retarded," was able to memorize books read to him after a single reading at the age of 18 months, and by age six had memorized the entire index of a set of encyclopedias?

Or if I told you that some children between the ages of two and five years old provide details of a "previous life"—some of which are historically verified, some of which involve the children speaking a foreign language that they were never taught, and some of whom are born with distinctive birthmarks or physical deformities that match medical records describing how the person died in the "previous life"?

Or if I told you that in controlled, more than double-blind studies conducted over the phone, certain people (known as "psychic mediums") are able to obtain specific, accurate information about a random person's deceased relative—claiming they communicate with dead people?

Or if I told you that by focusing one's mental attention on a "random number generator" machine that randomly produces 0's and 1's, ordinary people can affect the pattern of 0's and 1's ever so slightly without physically touching the machine, and they can do it from miles away?

And what if I told you that in many instances, these topics have been studied scientifically under controlled conditions over many decades, and the results have been published in peer-reviewed journals?

And what if I told you that some of the scientists studying these concepts are from Harvard University, Princeton University, Yale University, Stanford University, Cornell University, Cambridge University, Duke University, Johns Hopkins University, and the University of Virginia (among many others)?

You might say what most people have said to me: "I find all of that hard to believe. But real or not, it's fascinating!"

You might also be wondering if there is a connection between these seemingly unbelievable examples. There is, in fact, a connection. The connection is that they deal with "consciousness"—our mind, our inner experience and awareness, and our feeling of experiencing life. More specifically, the examples deal with the connection between the brain and consciousness. They suggest that consciousness is not produced by the brain and instead that consciousness exists independently of it. Using a crude analogy,[2] it's as if the brain is an antenna (such as those used in cell phones and televisions) that receives, transmits, processes, and filters signals that exist outside the body. In other words, our brain picks up consciousness from "the cloud." These ideas fly in the face of mainstream scientific thought, which typically regards consciousness as a by-product of brain activity.

If this is true, then one might wonder: Do we all have hidden, wizard-like abilities? What happens to our consciousness after we die? How does consciousness interact with the past, present, and future? What is the role of consciousness in the universe? What might this mean for science, philosophy, and human civilization?

These are the ideas and questions we will explore in this book.

About me

Books of this nature are typically written by scientists and philosophers. I am neither. Rather, I am a businessman. I began my career as an analyst in the New York office of a global investment bank. My first day of work was in July of 2008—right before the greatest financial crisis since the Great Depression. My clients were the very financial institutions that were under duress. I survived layoffs, which simply meant that I had to take on even more work. So I was in a distressed industry, working with distressed clients, at an under-resourced investment bank. Like most investment bankers during that period, I worked around the clock, pulling all-nighters, and spending many, many consecutive weekends in the office without a day off. Unfortunately, the Wall Street lifestyle didn't leave time for me to explore my latent scientific interests.

I always had nagging questions about the universe and our existence. When I started as an undergraduate at Princeton University in 2004, I spent my first year and a half focusing on economics courses. I also dabbled in physics. Physics, particularly astrophysics, thrilled me—it explored questions around the nature of reality and our existence. What is more fundamental than that? Midway through my sophomore year, I considered making a switch from the economics department to the astrophysics department. I was so serious about it that I met with the Astrophysics Department head to determine whether I could meet the course requirements as a late entrant. I was intrigued by the opportunity to learn about the universe from world-class physicists.

But there was a problem with my idea. I was a member (and later captain) of Princeton's varsity men's tennis team, a Division I athletic program. It dawned on me that my tennis obligations would make a late entry into the Astrophysics Department a near impossibility.

As a compromise, instead of studying the invisible forces that govern the universe, I decided to study the invisible forces that govern (and secretly bias) human judgment and decision-making. I chose the Psychology Department, emphasizing behavioral economics. My academic work culminated in a thesis that proposed a novel, dynamic model of Prospect Theory—Daniel Kahneman's Nobel Prize-winning idea on how people make decisions under risk—which contributed to my graduating *magna cum laude.*

But my existential questions remained unanswered. Shifting gears from Princeton to investment banking during the financial crisis only further buried them.

After leaving Wall Street in 2010, I joined a technology-focused investment bank and strategy firm, of which I am currently a senior member in Silicon Valley. Upon joining, I had more free time outside of work to learn about theoretical physics. Studying physics was a hobby, and the business world remained my focus.

I've made a few public appearances in a business context, having been quoted in *Bloomberg Businessweek* and elsewhere. I have also authored internationally published business articles. However, this book is my first public foray into the scientific domain.

Why go public with this book, outside of my professional area of focus? Because in the course of my casual scientific research outside the office, I stumbled across concepts about which I knew very little—concepts which, if real, would change the world. They would transform science and how we think about our existence. On a personal level, the ideas caused a radical worldview shift and ultimately changed my life.

When I mentioned the topics to intelligent, highly educated friends and family members, they were intrigued. A number of people told me that their lives were shifting as a result of our discussions. Their outlook on life became more positive, and their lives started to improve. Something about the way I framed the information was connecting with people.

It dawned on me that there was a need for a comprehensive summary, written for a general audience. It needed to be written by an outsider who hadn't been tainted by biases of mainstream science. It needed to be written by a nonscientist and a nonphilosopher—someone who simply examines evidence and is unconcerned with what the data suggests, even if the data leads to ideas considered unpopular or taboo by contemporary academia.

In the summer of 2017, I made the decision to put my thoughts on paper. This book is the result.

Learning of strange phenomena

Starting in the summer of 2016, I stumbled across podcasts[3] and literature discussing phenomena that are unexplainable by the science I was taught. The following is a summary of what I was hearing:

Experts often talked about questions relating to consciousness. They were claiming that consciousness does not come from the brain. Rather, they described consciousness as being "nonlocal" to the body.

Some of the individuals nonchalantly spoke of their "psychic abilities" and "extrasensory perception (ESP) abilities." They claimed people had the ability to send their minds to distant locations in the past, present, or future and accurately describe what is, was, or will be there ("remote viewing"—also known as clairvoyance); they had an ability to communicate with other people by using their minds alone ("telepathy"); they had an ability to know the future before it happened ("precognition"); and they had an ability to affect physical objects using their minds alone, without making physical contact ("psychokinesis"). Furthermore, they said *all* humans have these abilities, but most of us aren't tuned in to them. And they said even animals demonstrate psychic abilities. They claimed the effects are often subtle and below the level of our normal perception, so we don't regularly experience them unless we are trained or are naturally talented.

Some of the individuals I came across were scientists who agreed that the above phenomena are, in fact, real. They said these phenomena have been proven in government-sponsored programs and in controlled laboratory studies, most of which haven't garnered mainstream public attention, and many of which have been derided by the scientific community. Because the effects are typically subtle, repeated trials and statistics have been required.

Furthermore, the individuals I came across suggested that consciousness survives the death of the physical body. In other words: *We don't die.* They referenced studies on near-death experiences—cases in which people were clinically dead, and upon being resuscitated, recalled highly lucid memories suggestive of an afterlife. The experiences could not be explained as mere hallucinations caused by a dying brain because in some cases the brain was "off" during the time of the lucid experience. There was also discussion about "mediums" who claimed they could communicate with dead people. In a similar vein, reference was made to studies conducted for 50-plus years at the University of Virginia on more than 2,500 children who remember details of "previous lives." The memories could not be explained by ordinary means. Sometimes what the children recalled has been historically verified, and there is no known way that they could have learned that information.

The collective data suggested that consciousness is not localized to, or dependent upon, the brain. And further, it suggested that consciousness is primary in the universe—it is more fundamental than physical matter and exists beyond space and time. Along these lines, there was discussion that at the deepest level of reality, there is only one consciousness, and each of our individual minds is a part of it. This idea would explain why we experience the same physical world—not because the physical world exists independently of consciousness, but because we are all part of the same consciousness.[4] Nobel Prize-winning physicist Erwin Schrödinger even explicitly stated: "Multiplicity is only apparent, in truth, there is only one mind."[5]

The overall picture was of a reality that expanded beyond the physical, material world that we can see with our eyes. Pieces of it reminded me of what mystical traditions have been saying for millennia.

From confusion to intrigue

I was stunned, skeptical, and confused. *None* of these ideas conformed to what I was taught. Is mind-reading possible? Can we know the future? Can the mind alone interact with physical matter? Can we communicate with deceased loved ones? Are past lives real? Is consciousness fundamental in the universe and not a by-product of the brain? Are we all part of the same consciousness?

It would have been easy to dismiss any one person's testimony, which I certainly did on many occasions. *Charlatans exist in any domain*, I thought, *but that alone is not a reason to discount an entire field of study.* What struck me was how many of these individuals—who seemed intelligent and sane—were independently drawing the same picture. Why were they arriving at similar conclusions? Were they colluding behind the scenes? Were they all delusional? I needed to know more.

I did what anyone with my interests would do: I Googled the topics. Some of the first search results were from Wikipedia and other sources that were almost universally dismissive. For example, Wikipedia says about telepathy: "There is no convincing evidence that telepathy exists, and the topic is generally considered by the scientific community to be pseudoscience."[6] Some scientists agree. For example, Arizona State University physicist Lawrence Krauss states: "It's not controversial at all. There's no scientific evidence for extra-sensory perception."[7]

Statements such as those were not sufficient for me. I needed to see for myself what the research showed. So I did what I was trained to do academically and professionally: I read primary sources, I found recorded interviews with credible scientists, and I spoke with the experts myself. My research began to draw a very different picture than what Wikipedia and other sources said.

That opened Pandora's Box.

I learned of not just a few, but of *an abundance* of controlled studies in peer-reviewed journals suggesting that psychic phenomena (such as remote viewing, telepathy, precognition, and psychokinesis) are real.[8]

I learned, for example, that the statistical support of these phenomena is stronger than the evidence showing that aspirin prevents heart attacks.[9] And that an analysis of decades of such studies showed stronger statistical results than the highly publicized "proof" of the existence of the Higgs boson particle (a groundbreaking finding in physics that led to the winning of a Nobel Prize).[10] If the results for psychic phenomena are so strong, then why doesn't science accept that these abilities are real? If we accept similar statistical results in other areas of science, why don't we accept them here? And furthermore, if this stuff is real, do we have any science that might explain it?

I brushed up on my quantum physics, an area of science that has been proven for nearly a century but which is nothing short of bizarre. Albert Einstein famously called it "spooky." Quantum physics teaches us that the universe doesn't work in a way that aligns with our everyday experience. It teaches that there are seemingly invisible connections between objects that are physically distant. It teaches that the mere act of observing can affect physical reality. These ideas led Nobel Prize-winning physicist Brian Josephson to state in 2001: "Yes, I think telepathy exists…and I think quantum physics will help us understand its basic properties."[11]

So there exists science that *might* be able to explain the weird phenomena I learned of. This was a far cry from what my initial Internet research showed.

Resistance to new paradigms

But in my research I realized that mainstream science resists these ideas. It vehemently rejects the notion that consciousness can exist independently of the body. Mainstream science instead holds to its belief that consciousness is produced by the brain, without being able to explain how. It also assumes that consciousness has no effect on the physical world. If you pick up a cutting-edge physics book in a bookstore, the odds are that consciousness is either minimally referenced or completely absent.

Physicist Lee Smolin comments: "I get a lot of e-mails about consciousness. To most of them I reply that whereas there are real mysteries about consciousness, they're beyond what science can tackle with present knowledge. As a physicist, I have nothing to say about them."[12]

Furthermore, Dr. Eben Alexander, formerly a Harvard Medical School associate professor in brain surgery, states: "Depending on whom you talk to, consciousness is either the greatest mystery facing scientific enquiry, or a total nonproblem. What's surprising is just how many more scientists think it's the latter. For many—maybe most—scientists, consciousness isn't really worth worrying about because it is just a by-product of physical processes. Many scientists go further, saying that not only is consciousness a secondary phenomenon, but that in addition, it's not even *real*"[13] [emphasis in original].

For scientists who feel that way, the notion that consciousness might be vitally important to our understanding of the universe threatens to turn their worlds upside down. Resistance inevitably ensues. And the more I researched, the more this resistance became apparent. On the one hand, smart people claim there is strong scientific evidence that consciousness is not localized to the brain or body. On the other hand, a number of intelligent people claim there is no such evidence. And furthermore, they claim that if these phenomena are somehow real, then all of science would need to change.

One scientist who thinks there is strong evidence is psychologist Dean Radin, PhD, chief scientist at the Institute of Noetic Sciences (IONS). IONS is a Petaluma, California-based consciousness research institution that was founded in 1973 by Apollo 14 astronaut Edgar Mitchell. Dr. Radin has spent roughly 40 years conducting and examining studies on psychic phenomena and has held appointments at Princeton University and AT&T Bell Labs. He also conducted classified research on psychic phenomena for the U.S. government.[14]

In his 2018 book, *Real Magic*, he discusses categories of psychic phenomena that have achieved "six-sigma" statistical results under controlled experimental conditions. Six-sigma indicates that the effects are likely to be real: The odds that the results are occurring simply by chance, "after careful consideration of all known experiments investigating the same topic, are assessed to be *over a billion to one*"[15] [emphasis added].

Dr. Radin describes studies in the following areas that have achieved six-sigma results:

- Remote viewing[16]
- Telepathy[17]
- Precognition[18]
- Psychokinesis[19]

He comments: "Each of these experiments used protocols that avoided all known design flaws. An extensive due diligence of possible design faults has developed after years of intense scrutiny and criticism of these studies, leading to bulletproof designs. Each class of experiments has been repeated by a dozen to more than a hundred times by independent investigators at different labs around the world, with each class cumulatively involving hundreds to thousands of participants. The vast majority of the studies involved ordinary people, most of whom were not claiming any special [psychic] abilities."[20]

Lund University psychologist Dr. Etzel Cardeña finds similar results. In May 2018, his analysis of the evidence for psychic phenomena was published in *American Psychologist*, the official peer-reviewed academic journal of the American Psychological Association. The fact that these results have been published in such a mainstream journal is significant. As summarized by Dr. Cardeña: "The evidence provides cumulative support for the reality of [psychic phenomena], which cannot be readily explained away by the quality of the studies, fraud, selective reporting, experimental or analytical incompetence, or other frequent criticisms. The evidence...is comparable to that for established phenomena in psychology and other disciplines."[21]

A number of scientists who have examined the data agree with Drs. Radin and Cardeña. Others maintain that there is no evidence. The table on the next page illustrates the divided dynamic that we see in science.

Claims that psychic phenomena are real (examples)	**The other side (examples)**
"Using the standards applied to any other area of science, it is concluded that psychic functioning has been well established. The statistical results of the studies examined are far beyond what is expected by chance. Arguments that these results could be due to methodological flaws in the experiments are soundly refuted. Effects of similar magnitude to those found in government-sponsored research...have been replicated at a number of laboratories across the world. Such consistency cannot be readily explained by claims of flaws or fraud.... This is a robust effect that, were it not in such an unusual domain, would no longer be questioned by science as a real phenomenon."[22] — Jessica Utts, member of the CIA's review panel for experiments on remote viewing run at Stanford University; she was also the 2016 president of the American Statistics Association (1995)	"It's perfectly acceptable to test psychic and paranormal phenomena like ESP...and in fact those tests have been done. But they always fail."[23] — Jerry Coyne, University of Chicago professor, Department of Ecology and Evolution Sciences (2014)
"With regard to [psychic] phenomena, here I will simply say...the thousands of field and laboratory studies carried out by competent scientists over the 130-plus years since the founding of the Society for Psychical Research cumulatively provide an overwhelming body of evidence—*for those who will take the trouble to study it with an open mind*—that these phenomena really do exist as facts of nature"[24] [emphasis in original]. — Ed Kelly, University of Virginia professor of psychiatry and Neurobehavioral Sciences (2015)	"Psychic powers...don't exist. We can say that with confidence, even without digging into any controversies about this or that academic study. The reason is simple: what we know about the laws of physics is sufficient to rule out the possibility of true psychic powers."[25] — Sean Carroll, California Institute of Technology professor of physics (2016)

Claims that psychic phenomena are real (examples)	**The other side (examples)**
"Yes, I think telepathy exists…and I think quantum physics will help us understand its basic properties."[26] — Brian Josephson, Nobel Prize-winning physicist (2001)	"[Physicist Sean Carroll] deftly shows how current physics is so solid that it rules out ESP forever."[27] — Steven Pinker, Harvard professor of psychology (2016)
"It appears quite clear…that irrespective of what interpretation is given to specific research reports, the overall results of [psychic] experimentation are indicative of an anomalous process of information transfer, and they are not marginal and neither are they impossible to replicate. In the face of this, the critic who merely goes on asserting there is no evidence for [psychic phenomena] is using a tactic reminiscent of Mohammed Saeed al-Sahaf, Iraq's former information Minister, in blindly asserting there are no American troops in Baghdad."[28] — Adrian Parker, Goteborg University professor of psychology; and Göran Brusewitz of the Swedish Society for Psychical Research (2003)	"It's not controversial at all. There's no scientific evidence for extra-sensory perception."[29] — Lawrence Krauss, Arizona State University physicist (2017)
"The empirical evidence for the non-local nature of consciousness emerging from the research at [Princeton Engineering Anomalies Research Lab (PEAR)] and elsewhere inescapably predicates questions about the non-physical…dimensions of human experience."[30] — Robert Jahn, former dean of engineering at Princeton University, and PEAR Laboratory Manager Brenda Dunne (2011)	In response to a study on precognition: "If any of his claims were true, then all of the bases underlying contemporary science would be toppled, and we would have to rethink everything about the nature of the universe."[31] — Douglas Hofstadter, cognitive scientist at Indiana University (2011)

Claims that psychic phenomena are real (examples)	The other side (examples)
"I assume that the reader is familiar with the idea of extrasensory perception, and the meaning of the four items of it, viz., telepathy, clairvoyance, precognition and psychokinesis. These disturbing phenomena seem to deny all our usual scientific ideas. How we should like to discredit them! Unfortunately the statistical evidence, at least for telepathy, is overwhelming."[32] — Alan Turing, pioneering computer scientist who helped to crack German codes in World War II (1950)	If telepathy occurred, it would "turn the laws of physics upside down."[33] — Richard Dawkins, evolutionary biologist formerly of Oxford University
"I never liked to get into debates with skeptics, because if you didn't believe that remote viewing was real, you hadn't done your homework."[34] — Major General Edmund R. Thompson, Army Assistant Chief of Staff for Intelligence, 1977–1981, and Deputy Director for Management and Operations, DIA, 1982–1984	Proof of remote viewing "would overturn almost everything we know in science."[35] — Ray Hyman, psychologist and professor emeritus at the University of Oregon (2002)
"Unless there is a gigantic conspiracy involving some thirty University departments all over the world, and several hundred highly respected scientists in various fields, many of them originally hostile to the claims of psychical researchers, the only conclusion the unbiased observer can come to must be that there does exist a small number of people who obtain knowledge existing in other people's minds, or in the outer world, by means as yet unknown to science."[36] — Professor H.J. Eysenck, chairman of the Psychology Department, University of London (1957)	Regarding psychic phenomena: "Any confirmation, *no matter how weak an effect*, would force a radical change in our worldview" [emphasis in original].[37] — Bruce Rosenblum and Fred Kuttner, physicists at UC Santa Cruz (2011)

Can you sense the tension?

Dr. William Tiller, former head of the Material Sciences Department at Stanford University, sums up the situation well. He states that mainstream materialist science has "known about this category of...psychic phenomena for one to two centuries, but...since it doesn't satisfy their internal self-consistency requirements of orthodox science, they either have to change their attitudes and their way of doing experiments or they have to sweep it under the rug. Unfortunately they've chosen to sweep it under the rug....They're terribly stuck....They're kind of afraid, I think, to get out of the box....Universities are the same; universities won't allow this stuff to be taught, to get a PhD degree in this."[38]

This dynamic is apparent in the research community. Psychologist Imants Barušs and cognitive neuroscientist Julia Mossbridge comment: "The results of research concerning anomalous phenomena are often unjustly treated...and papers in which the occurrence of anomalous phenomena are reported are sometimes rejected for publication in mainstream journals irrespective of their quality."[39] Cambridge's Andreas Sommer makes a similar claim: "Whilst critics continue to launch uninformed but widely publicized attacks, editors of mainstream journals have admitted to stick to the rule of rejecting papers reporting positive [psychic] effects irrespective of the quality of submitted manuscripts."[40]

Science has a habit of getting it wrong before getting it right

If consciousness really does exist outside the brain, then science would need to shift its paradigms. But paradigm shifts threaten mainstream thought. Many smart scientists would have to admit that their theories are incomplete or wrong.

We've seen this movie before, though. Throughout the annals of history, people have held on to beliefs, thinking they knew everything, until they realized they didn't—and this has happened over and over and over again. Many ideas that are now commonly accepted were originally dismissed and ridiculed.

We used to think that the earth is flat. Galileo was convicted of heresy for teaching that the earth isn't the center of the universe and that it revolves around the sun. We used to ridicule scientists claiming that invisible germs could harm us (until we could see them through advanced microscopes). The list could go on and on. It's easy in hindsight to criticize what now seems obvious. Could the assumption that "the brain produces consciousness" be next on this list? Dr. Tiller, among others, thinks so. He calls the shift a "Copernican-scale Revolution."[41] In other words, he thinks we're on the cusp of the next scientific revolution—a drastic shift in thinking about basic assumptions we've held for so long.

But how could our current paradigms be *so* far off? Is it possible that many brilliant scientists are missing something *that* big? It seems like science has come so far.

Science has indeed come a long way. However, we should remember how much remains unknown. Physicists have identified only ~4 percent of the matter that makes up the universe. The remaining 96 percent is labeled "dark matter" and "dark energy." We know something is there, but we don't know what it is—and it's the vast majority of the universe! As stated by Jim Peebles, professor emeritus of physics at Princeton University: "It is an embarrassment that the dominant forms of matter in the universe are hypothetical."[42]

It's perhaps not surprising, then, that mainstream physics has been unable to devise a unifying "theory of everything" to explain the universe. Instead, we have individual theories, such as quantum mechanics and general relativity, which work in specific cases but are incompatible when applied together. So, basically, we have no idea how the universe works.

Instead of humbly reminding ourselves how little we know, there is a tendency to think we know it all. That hasn't worked out well in the past. In 1894, Nobel Prize-winning physicist Albert Michelson proclaimed: "The more important fundamental laws and facts of physical science have all been discovered, and these are now so firmly established that the possibility of their ever being supplanted in consequence of new discoveries is exceedingly remote.... Our future discoveries must be looked for in the sixth place of decimals."[43]

However, in 1900, British physicist Lord Kelvin noted that while physics had most of the answers, "two clouds"[44]—two mysteries—remained that physics couldn't explain. Well, they were soon explained. And they led to the discoveries of general relativity and quantum mechanics—two of the most revolutionary theories in the history of science.

These prominent and brilliant physicists thought they knew it all, but they couldn't have been more wrong.

Could the anomalies of nonlocal consciousness: psychic abilities, ESP, near-death experiences, etc., be the "clouds" of today's science—that is, "clouds" that we'd prefer to sweep under the rug because they don't align with our theories, but which could ultimately lead to a new understanding of human capabilities? Are we just like the scientists at the turn of the 20th century who thought they knew it all? Are we quick to dismiss phenomena just because they seem too unbelievable to be true?

"Have you looked at the evidence?"

Dr. Jessica Utts, the 2016 president of the American Statistics Association, has studied the data on psychic research. She concludes, based on statistical analysis, that these phenomena are real. She even made this declaration in a 1995 report that she created at the request of Congress and the CIA.[45]

She makes an important general observation that is worth remembering as you read this book: "It is too often the case that people on both sides of the question debate the existence of psychic functioning *on the basis of their personal belief systems rather than on an examination of the scientific data*"[46] [emphasis added].

Science should be governed by an unemotional examination of evidence. It should not be based on what we "want" or "believe" to be true. Nor should it be governed by what we don't want to be true. As stated by physicist Neil deGrasse Tyson: "The good thing about Science is that it's true whether or not you believe in it."[47]

However, Dr. Utts observes unscientific behavior among scientists: "Most scientists reject the possible reality of [psychic] abilities without ever looking at data!...I have asked the debunkers if there is any amount of data that could convince them, and they generally have responded by saying, 'probably not.' I ask them what original research they have read, and they mostly admit that they haven't read any!"[48]

For example, remote viewing researcher Stephan A. Schwartz recalls a debate with a skeptic:

> Along with [physicist] Ed May, I once debated with [Tufts professor] Daniel Dennett, a prominent critic of ESP research, at an event produced by ABC News for station news staffs and station managers. We debated along for about thirty minutes, with Dennett making dismissive and disparaging remarks

> to anything Ed or I said, but always in generalities. Finally I said to him: "Let's pick an experiment we both know, and you tell me what is wrong with it, and I will respond." Without a moment's hesitation he shot back...saying, "You don't think I actually read this stuff, do you?"...It suddenly dawned on Dennett what he had said. He blushed and sat down, and left as soon as he could.[49]

In another case, Nobel Prize-winning physicist Brian Josephson included a controversial comment in a 2004 booklet accompanying the hundredth anniversary of the Nobel Prize. He remarked, "Quantum theory is now being fruitfully combined with theories of information and computation. These developments may lead to an explanation of processes still not understood within conventional science such as telepathy."

Josephson's acknowledgment that telepathy might be real evoked fury from mainstream scientists, including Oxford physicist David Deutsch. Deutsch responded to Josephson's comment: "It is utter rubbish... Telepathy simply does not exist. The Royal Mail has let itself be hoodwinked into supporting ideas that are complete nonsense." However, Josephson commented: "It may be relevant to note that Deutsch, has never, as yet, rsponded [sic] to any emails people have sent him asking how much he has actually studied the...literature."[50]

In yet another case, former Cambridge biochemist Rupert Sheldrake, PhD, squared off against the often-skeptical former Oxford biologist Richard Dawkins. Sheldrake was asked to speak with Dawkins shortly before Dawkins's then-upcoming show *The Enemies of Reason* (released in 2007). Sheldrake was hesitant to partake because he worried that the show would be biased. But he agreed to participate, stating: "On the understanding that Dawkins was interested in discussing evidence, and with the written assurance that the material would be edited fairly, I agreed to meet him and we fixed a date."

Sheldrake and Dawkins debated. As Sheldrake recalls, Dawkins's view was that "in a romantic spirit he himself [Dawkins] would like to believe in telepathy, but there just wasn't any evidence for it. [Dawkins] dismissed all research on the subject out of hand, without going into any details."[51] After debating further, Sheldrake and Dawkins agreed that in order to test whether telepathy and other such phenomena are real, they needed to conduct controlled experiments.

Sheldrake then recalls:

> I said [to Dawkins] that this is exactly why I had actually been doing such experiments, including tests to find out if people really could tell who was calling them on the telephone when the caller was selected at random. The results were far above the chance level. The previous week I had sent Dawkins copies of some of my papers in scientific journals so that he could examine some of the data before we met. At this stage he looked uneasy and said, "I don't want to discuss the evidence." "Why not?" I asked. He replied, "There isn't time. It's too complicated. And that's not what this program is about." The camera stopped.[52]

These examples are eerily similar to what Galileo faced centuries ago when his world-changing evidence challenged the mainstream. He claimed that the earth isn't the center of the solar system, and instead, the earth revolves around the sun. Galileo's idea was highly controversial and went against "common sense." People saw the sun move across the sky each day, so they assumed it was obvious that the sun revolved around the earth. We now know Galileo's theories to be true and regard the alternative as silly.

Sarah Knox, biomedical science professor at West Virginia University Medical School, likens the struggles of today's nonlocal consciousness researchers to Galileo's struggles centuries ago: "Some [critics contend] because there is no plausible mechanism within a materialist frame of reference to explain them, paranormal phenomena can't possibly be valid. *This is the same reasoning that the learned men of Galileo's day used when they refused to look in the telescope.* This attitude is nowhere more evident than in the number of scientists who are willing to volunteer as 'expert' commentators on television programs about paranormal phenomena, astonishingly undeterred and unembarrassed by their complete lack of knowledge concerning the experimental data"[53] [emphasis added].

But what happens when one does look in the telescope?

For most of astronomer Carl Sagan's career, he was skeptical of the idea that consciousness could exist outside the brain, arguing (like many mainstream materialist scientists) that consciousness arises from brain activity: "[The brain's] workings—what we sometimes call mind—are a consequence of its anatomy and physiology, and nothing more."[54]

In 1994, Sagan was speaking with psychologist Daryl Bem, PhD, his colleague at Cornell. Dr. Bem had studied ESP and told Sagan about his work. Sagan's response seemed to suggest that there are "no replicable findings"[55] in these areas. When Dr. Bem asked if Sagan had actually reviewed the findings, Sagan said he had not. Dr. Bem then sent Sagan some of his recent scientific research and asked that Sagan review the findings before making dismissive comments.

Apparently, the paper was impactful. In Sagan's last published book before he died, *The Demon-Haunted World: Science as a Candle in the Dark* (1996), he stated: "There are three claims in the ESP field which, in my opinion, deserve serious study: (1) that by thought alone humans can (barely) affect random number generators in computers; (2) that people under mild sensory deprivation can receive thoughts or images 'projected' at them; and (3) that young children sometimes report details of a previous life, which upon checking turn out to be accurate and which they could not have known about in any other way than reincarnation."[56]

Sagan peeked in the telescope, and you see what happened.

Want to look in the telescope with me?

This book is your peek into the telescope—a look at the evidence. We will examine the concepts that intrigued Carl Sagan, and more. I have summarized a large body of evidence so that your peek into the telescope will be meaningful and time-efficient—that is, if you want to look. In many cases I reference scientific studies, but I merely summarize them. For the purposes of this book, I do not dwell on methodology beyond an overview. Those who feel an urge to read the primary sources can review the endnotes and bibliography for more information.

Section II of the book lays the foundation before diving into scientific evidence. We will explore the connection between the brain and consciousness. We will learn that—counter to what many of us were taught—science does *not* know where consciousness comes from. We will examine cases in which *reduced* brain functioning is related to *heightened* conscious experience. This relationship might make sense if we regard the brain as a filter—rather than a producer—of consciousness. Next we will examine, in a simplified fashion, accepted science—quantum physics, relativity theory, and chaos theory—which is important to understand before we dive into the phenomena described later in the book. These areas of science can teach us that reality sometimes works in counterintuitive ways. And therefore, the strange phenomena later described might be plausible in the context of a counterintuitive universe.

Section III delves into phenomena suggesting that we all have psychic abilities. We will review scientific findings on remote viewing, telepathy, precognition, psychic animals, and psychokinesis.

Section IV examines whether consciousness survives the death of the physical body. We will review the science of near-death experiences, communications with the deceased, and children who remember "previous lives."

Section V explores implications. How is it possible that mainstream science is missing something so big? What does it mean for everyday life? How does Elon Musk's latest brain-focused endeavor fit in? What are the implications for artificial intelligence, medicine, safety systems, and national and personal security? What are the implications for how we think about life, death, and meaning...as well as love, beauty, happiness, and world peace? Ideas are presented in summary form, but the topics themselves could be (and have been) the subjects of books on their own.

My conclusion after examining the accumulated evidence

I conclude that it is highly unlikely that *every* example of nonlocal consciousness described in this book and elsewhere has been fabricated or misinterpreted. In other words, I think it is likely that *at least one* of them (if not more) is real. That means that materialism needs to be reconsidered. Alternatively, a framework in which consciousness is the basis of reality explains the phenomena well. And if that framework is correct, we are on the brink of perhaps the most significant revolution in human history.

To summarize, if consciousness is fundamental, the key implications are:

- Materialism—the foundational assumption of modern science and much of modern thought—is wrong.
- Consciousness is not produced by the brain; rather, consciousness is "nonlocal" to the brain/body system.
- We all have latent psychic abilities.
- When our body dies, our consciousness does not die.

- Consciousness exists beyond space and time.
- We are all fundamentally interconnected as part of the same underlying consciousness.

How might my views and writing style impact your experience of the book?

My research has convinced me that a paradigm shift away from materialism is warranted. That bias inevitably imbues my writing. But regardless of how convincing I personally find the evidence to be, my hope is that this book will serve as a guide to help you navigate alternative perspectives on our reality. I'm simply bringing together the existing evidence for you, however strong or weak you may think it is.

I attempt to simplify technical topics as much as possible while retaining the core meaning. My goal in doing so is to make the prose accessible to a wide variety of readers. The simplified bullet-point chapter summaries at the end of chapters 2 through 11 and the glossary of terms can further assist you.

Why is this book important?

My hope is that this book will provide a spark to advance the conversation throughout society—in science and in everyday life. Unfortunately, we live during a time in which studying these topics is too often discouraged by the mainstream.

For example, the former editor of *Nature*, Sir John Maddox, called Dr. Rupert Sheldrake's 1981 book *A New Science of Life* "the best candidate for burning there has been for many years." Maddox then said in a BBC interview in 1994: "Sheldrake is putting forward magic instead of science, and that can be condemned in exactly the language that the Pope used to condemn Galileo, and for the same reason. It is heresy."[57]

Dr. Sheldrake isn't the only scientist to face these issues. Drs. Barušs and Mossbridge sum up the disturbing state of affairs: "As a result of studying anomalous phenomena or challenging materialism, scientists may have been ridiculed for doing their work, been prohibited from supervising student theses, been unable to obtain funding from traditional funding sources, been unable to get papers published in mainstream journals, had their teachings censored, been barred from promotions, and been threatened with removal from tenured positions. Students have reported being afraid to be associated with research into anomalous phenomena for fear of jeopardizing their academic careers. Other students have reported explicit reprisals for questioning materialism, and so on."[58]

For example, Dr. Mossbridge comments: "In my own experience, I have been advised multiple times not to put my [precognition] research on my resume, and when I have tried to publish [precognition] results in the mainstream literature, I have been told I was risking my scientific career."[59]

This needs to end. It's time to put real scientific effort into exploring anomalies of consciousness. It's time to start funding research to build the evidence base and to augment or refute theories. It's time for us to look in the telescope in earnest.

As Nikola Tesla is quoted as saying: "The day science begins to study non-physical phenomena, it will make more progress in one decade than in all the previous centuries of its existence."[60]

Or as Dr. Robert Jahn and Brenda Dunne of Princeton's Engineering Anomalies Research Lab put it: "It is our belief that here we are teetering on the threshold of another new era of science…that will acknowledge and utilize sublime capabilities of the proactive human mind to extract far deeper aspects of physical experience."[61]

And moreover, Dr. Tiller of Stanford states: "For the last four hundred years, an unstated assumption of science is that human intention cannot affect what we call 'physical reality.' Our experimental research of the past decade shows that, for today's world and under the right conditions, this assumption is no longer correct. We humans are much more than we think we are."[62]

We're now ready to look in the telescope.

Section II

Laying the Foundation

This section discusses two primary topics: the relationship between the brain and consciousness; and the science of counterintuitive, but accepted, physics. It is important to have an understanding of these topics before delving into the phenomena discussed in sections III and IV.

Chapter 2

The Unproven Assumption
"The Brain Creates Consciousness"

Science knows surprisingly little about mind and consciousness. Current orthodoxy holds that consciousness is created by electrochemical reactions in the brain, and that mental experiences fulfill some essential data-processing function. However, nobody has any idea how a congeries of biochemical reactions and electrical currents in the brain creates the subjective experience of pain, anger, or love.... We have no explanation and we had better be clear about that.[1]

—Yuval Noah Harari, author of *Sapiens* and *Homo Deus*

There is nothing about a brain, studied at any scale, that even suggests *that it might harbor consciousness*[2] [emphasis in original].

—Sam Harris, neuroscientist

We can't even begin to explain how consciousness, how sensation, arises out of electric chemistry.[3]

—Henry Marsh, neurosurgeon

It's absurd! Scientists have yet to explain the nature of consciousness. They have no means of objectively detecting it. They have not identified its necessary and sufficient causes. And yet, they ask us to wager everything on their belief that consciousness is solely a product of the brain.[4]

—B. Allan Wallace, Buddhist philosopher

> *Nothing in modern physics explains how a group of molecules in your brain create consciousness. The beauty of a sunset, the miracle of falling in love, the taste of a delicious meal—these are all mysteries to modern science. Nothing in science can explain how consciousness arose from matter. Our current model simply does not allow for consciousness, and our understanding of this most basic phenomenon of our existence is virtually nil. Interestingly, our present model of physics does not even recognize this as a problem.*[5]
>
> —Stem-cell biologist Robert Lanza, MD; and physicist Bob Berman, in their book, *Biocentrism*

As discussed in the preface, we don't know where consciousness comes from. That fact alone might blow your mind. It certainly did for me when I first learned of this notion. I had always been taught to assume that the brain is responsible for producing my conscious experience. But the fact is: We don't know how physical, seemingly unconscious matter creates a nonphysical consciousness.

In general, science has made immense strides throughout history. It has enabled us to travel to the moon, to build smartphones, to genetically modify biological organisms, and much more. Yet, in spite of our progress, we still don't know where our mind comes from! The most undeniable and obvious part of our existence is that we have the subjective, inner experience of feeling alive, and we still can't explain it.

In this section we will elaborate on this critical topic and further explore the question that materialism can't seem to answer: "Does the brain produce consciousness?"

Defining consciousness

In the preface and introduction, I loosely defined *consciousness* as the mind, an inner experience and awareness. One could use the following example as a guide: Your consciousness is experiencing the reading of these words right now. Consciousness is your sense of being you, and it is your sense of experiencing life. When you say, "*I* am reading this book," you could regard "I" as consciousness.

While I will use the above definition for the purposes of this book, it is important to note that different people have different definitions of consciousness. For instance, as summarized in her 2012 book, *Consciousness:*

Bridging the Gap Between Conventional Science and the New Super Science of Quantum Mechanics, Eva Herr interviewed ten prominent scientists and philosophers and found that they each had different definitions of consciousness.

That shouldn't stop us from using the above definition, but we should simply acknowledge the definitional question that remains.[6]

The hard problem of consciousness

Now that we have settled on a general definition, let's further explore the controversy. We were taught (perhaps implicitly) in biology class that the brain is what is responsible for our consciousness. That idea is so ingrained in our culture that we might not even realize that we make an assumption.

Why do we assume that the brain produces consciousness? Physicist Peter Russell provides one theory:

> Our primary senses, our eyes and ears, happen to be situated on the head. Thus the central point of our perception, the point from which we seem to be experiencing the world, is somewhere behind the eyes and between the ears—somewhere, that is, in the middle of the head. The fact that our brains are also in our heads is just a coincidence, as the following thought experiment bears out. Imagine that your eyes and ears were transplanted to your knees, so that you now observe the world from this new vantage point. Where would you now experience your self to be—in your head or down by your knees? Your brain may still be in your head, but your head is no longer the central point of your perception. You would now be looking out onto the world from a different point, and you might well imagine your consciousness to be in your knees.[7]

Another reason why we might assume that the brain produces consciousness is that there is a strong correlation between brain activity and conscious experience. But that still doesn't explain how the brain *produces* consciousness. As stated by biochemist Dr. Rupert Sheldrake: "Even if we understand how eyes and brains respond to red light, the *experience* of redness is not accounted for"[8] by our current understanding of the brain [emphasis in original].

Let's assume for a moment that materialism is correct and that the brain produces conscious experience. Let's explore the miracle that this would imply. Think about your thoughts and feelings. You know you are experiencing them, but you can't touch them. They are nonphysical. How is it that these nonphysical thoughts and feelings of the mind magically arise from the physical matter of the brain? How is it that trillions of cells in the human body come together in a way that allows nonphysical conscious experience to emerge? This is precisely what makes the "hard problem" of consciousness so difficult to solve.

We don't have an answer. As stated by philosopher Alva Noë: "After decades of concerted effort on the part of neuroscientists, psychologists, and philosophers, only one proposition about how the brain makes us conscious—how it gives rise to sensation, feeling, subjectivity—has emerged unchallenged: we don't have a clue."[9]

We can't rely on an assumption that neuroscience will someday provide the answer. Should we simply wait for an answer that may never come?

Even Nobel laureate Francis Crick looked at this issue. Many know of Crick as the brilliant scientist who changed the world by codiscovering the double helix structure of DNA with James Watson. Few know that he then devoted the rest of his life to trying to prove that the brain produces consciousness. You don't hear as much about that because he failed.[10] But so has every other scientist who has tried.

Instead of assuming that "the brain creates consciousness; we just don't understand yet how it happens, but one day we will," we can consider an alternative. Perhaps the brain does not produce consciousness, and that is why we can't answer the hard problem of consciousness. Perhaps consciousness exists independently of the brain (and the body) and the brain is merely a filtering mechanism—a localization process—for consciousness.

Larry Dossey, MD, summarizes this stance in the following quotation from his 2013 book, *One Mind*:

> There are many reasons why scientists have assumed that the mind and brain are one and the same. When the brain is damaged through physical trauma or stroke, mental function can be deranged as a result. Vitamin deficiencies and malnutrition can cause impairment of thought processes, as can various environmental toxins. Brain tumors and infections

> can wreak havoc with mentation. In view of these effects, it has seemed reasonable to assume that mind and brain are essentially identical. But none of these observations *prove* that the brain produces the mind or that the mind is confined to the brain. Consider your television set. Although you can damage it physically and destroy the picture on the screen, this does not prove that the TV set actually makes the picture. We know, rather, that the picture is due to electromagnetic signals originating outside the set itself and that the TV set receives, amplifies, and displays the signals; it does not produce them. All we ever observe is the concomitant variations or correlations between states of the brain and states of the mind.... [Consider the] venerable maxim of science that "correlation is not causation." Night always follows day; the correlation is 100 percent; but that does not mean that day causes night[11] [emphasis in original].

Similarly, Gary Schwartz, who earned his PhD in psychology from Harvard, served as a professor at Yale and is currently a professor at the University of Arizona, agrees with Dr. Dossey's stance. After spending most of his career holding the materialist view of "brain first, mind second," Dr. Schwartz now concludes: "Mind is first. Consciousness exists independently of brain activity. It does not depend upon the brain for its survival. Mind is first, the brain is second. The brain is not the creator of the mind, it is a powerful tool of the mind. The brain is an antenna/receiver for the mind, like a sophisticated television or cell phone."[12]

Diane Powell, a Johns Hopkins MD, former Harvard Medical School faculty member, and practicing neuropsychiatrist, has come to a similar conclusion. She thinks materialist neuroscience is on the wrong path by focusing on the brain when trying to understand consciousness: "Trying to understand consciousness by investigating the gray matter in our skulls is like trying to comprehend music by dismantling CD players and analyzing their parts."[13]

Put another way, she compares materialist neuroscience's efforts to "looking...at the hardware and thinking you're going to understand the software."[14]

Eben Alexander, who earned his MD from Duke University and served on the faculty at Harvard Medical School as an associate professor in brain surgery, holds the same view. He says, "As a neurosurgeon, I was taught that the brain creates consciousness.... The truth is that the more we come to understand the physical brain, the more we realize it does not create consciousness at all. We are conscious in spite of our brain! The brain serves more as a reducing valve or filter, limiting pre-existing consciousness down to the trickle of the illusory 'here-now.'"[15]

And British psychologist Sir Cyril Burt stated: "Why should we assume that consciousness needs a material brain to produce it? A closer scrutiny of the actual facts makes it far more probable that the brain is an organ for selecting and transmitting consciousness rather than for generating it."[16]

The list of similar quotations could go on and on. In fact, in 2014 a group of more than 200 scientists and philosophers published *A Manifesto for a Post-Materialist Science*,[17] which takes a similar position.

The point is: There are many well-educated people who challenge the unproven materialist assumption that "the brain produces consciousness." You might not have known that until now. If there is anything you should take away from this chapter (and perhaps even from this book), it should be that we do not know where our mind comes from!

Let's now explore several examples that support the idea that materialism needs to be questioned—examples in which less brain activity translates into heightened or enriched conscious experience.[18]

Enriched realities: psychedelics and near-death experiences

Psychedelic substances such as psilocybin ("magic") mushrooms, LSD, ayahuasca, and others have been gaining attention in the news recently. For years, researchers were prohibited from using psychedelics in their experiments, but studies are again under way. The studies reveal potentially important findings related to consciousness.

Psychedelics produce an altered state of consciousness in which the user experiences a hyper-real reality that can be difficult to describe with words. Under the materialist view of consciousness, we might expect that these enriched experiences are caused by an increase in brain activity. A richer experience would be caused by more brain activity, right?

Well, that wouldn't be the case if the brain is a filter—a limiter—of some broader consciousness outside the body. If this is true, then we might expect psychedelics to *reduce* brain activity. For this reason, in 1954 novelist Aldous Huxley theorized, after experimenting with psychedelics, that they open the "reducing valve" of the brain.[19] In other words, this theory suggests that the brain normally limits our perception to a narrow view of reality, and psychedelics are a means of unlocking the filter and exposing us to the broader reality.

Nature reported on this idea in its 2012 article *Psychedelic chemical subdues brain activity*. The article profiles a study conducted by Carhart-Harris et al. (2012), who measured brain activity in two groups of participants: one injected with psilocybin—the hallucinogenic chemical found in magic mushrooms—and the other injected with a placebo. Those who took psilocybin reported a typical psychedelic experience. For instance, they reported experiencing their surroundings change in unusual ways, seeing geometric patterns, feeling unusual bodily sensations, having vivid imaginings, sensing an altered perception of size and space and time, hearing sounds that influenced their thoughts, and being in a dreamlike state. Participants given the placebo did not have these experiences.

In spite of the vivid experiences reported by psilocybin participants, the researchers commented, "We observed no increases in CBF or BOLD signal in any region."[20] (CBF stands for "cerebral blood flow," and BOLD stands for "blood-oxygen level dependent imaging.")

The materialist view of consciousness might have predicted that we would have seen an increase in these metrics, and yet we do not. Not only do the metrics not increase, but they instead decrease: Participants' brain images showed *reduced* brain activity compared to participants who took the placebo. As stated by the researchers: "These results strongly imply that the subjective effects of psychedelic drugs are caused by decreased activity and connectivity in the brain's key connector hubs, enabling a state of unconstrained cognition."[21]

University of Virginia professor of psychiatry and neurobehavioral sciences Dr. Ed Kelly, and University of California, Berkeley professor of molecular and cell biology Dr. David Presti, explain another significant finding: "The intensity of the psychedelic experience was significantly correlated with the magnitude of these decreases."[22] In other words, the psychedelic experiences tended to be *more* intense when brain activity was more greatly reduced.

Similarly, as we will examine in chapter 9, people who have "near-death experiences" have reduced or no brain function accompanied by lucid, hyper-real experiences that they struggle to put into words. The accounts resemble what we see in certain psychedelic accounts. Some near-death experiences have been reported by patients under general anesthesia. Other accounts are from patients who survived cardiac arrest during which there was no measured brain function. One study conducted by University of Virginia professors Emily Kelly, Bruce Greyson, and Ed Kelly found that 45 percent of near-death experiencers stated that their experiences were "clearer than usual," and 29 percent said they were "more logical than usual."[23] One has to wonder: Are the altered states of consciousness experienced in psychedelic trips and near-death experiences hallucinations? Or are they exposing us to some version of the "real" reality that is normally hidden by our limited brain? We will come back to this in chapter 9.

The takeaway point here is: If the brain is a "reducing valve" that filters and restricts us from experiencing a broader consciousness, then the above examples make sense. If the brain generates consciousness, as materialism assumes, then the above examples are "paranormal" and "anomalous."

Terminal lucidity

Another example is a mysterious phenomenon known as "terminal lucidity." The term describes "the unexpected return of mental clarity and memory shortly before death in patients suffering from severe psychiatric and neurologic disorders."[24] The return of clarity is reported to occur typically minutes, hours, or days before a person's imminent death. Reports of terminal lucidity have occurred in patients with Alzheimer's disease and schizophrenia; but they also occur in patients with brain abscesses, tumors, meningitis, strokes, and affective disorders. In other words, these are unexplained lucid experiences with impaired brains.

The number of cases reported formally throughout the last 250 years has been sparse: As of a 2011 study conducted by researchers at the University of Virginia and the University of Iceland, 83 cases had been documented. But the number of formally documented cases seems to understate how frequently terminal lucidity occurs. The study concluded that "*seven out of ten* caregivers in a nursing home reported that they had observed patients with dementia and confusion becoming lucid a few days before death during the past five years" [emphasis added]. The study further states:

"Interviewees from all [nursing home] units reported first-hand accounts of previously confused residents suddenly becoming lucid enough in the last days of life to recognize and say farewell to relatives and carers."

For example, consider the case of an elderly woman who had Alzheimer's disease for a decade and a half. She was unresponsive for many years and displayed no signs of recognizing her daughter (who cared for her) or anyone else. However, several minutes before her death, she began having a normal conversation with her daughter, who was "unprepared and... utterly confused" about what had occurred.

In another similar case, a woman who had Alzheimer's disease for years suddenly began speaking to her granddaughter and gave her life advice. The granddaughter said, "It was like talking to Rip Van Winkle."

Why are these people exhibiting unexpected *lucid* cognition with *impaired* brain function?

The researchers are similarly perplexed: "The unexpected return of mental faculties raises questions about cognitive processing at the end of life, especially in diseases that involve the degeneration of the brain regions usually responsible for complex cognition."[25]

Psychologist Dr. Imants Barušs and cognitive neuroscientist Dr. Julia Mossbridge make a similar point: "In neuroscience, we generally assume that mental clarity requires a functioning brain. Thus, the reason that mental clarity is unexpected in some of these [terminal lucidity] cases is the presence of obvious functional or structural brain pathology...that seems sufficient to *prohibit* mental clarity"[26] [emphasis added].

This phenomenon is certainly difficult to explain if we assume that the brain produces consciousness.

Savant syndrome

People with savant syndrome have profound mental abilities, but simultaneously have severe brain impairments. Sound familiar? Materialism also struggles to explain how this can happen.

Savant syndrome became popularized in the 1988 Oscar-winning movie *Rain Man*, a film based on the life of Kim Peek, a mentally retarded, but brilliant, man. Peek astonished *Rain Man* producer Barry Morrow when he knew "[his] date of birth and day of the week [he] was born, the day of the week this year, and day of the week and year [he] would turn 65 so [he] could think about retiring." Peek was also "familiar with almost every author and book in the library, quoting an unending amount of sports trivia, relating complex driving instructions to most anywhere,"[27] and was able to simultaneously read two different books—one with each eye.[28]

Dr. Darold Treffert, clinical professor of psychiatry at the University of Wisconsin School of Medicine, studied Peek extensively. In his 2010 book *Islands of Genius*, Dr. Treffert reports that "by age 18 months [Peek] was able to memorize all the books read to him with just a single reading.... By age six [Peek] was reciting whole paragraphs verbatim from a book with the mere mention of its page number. By that time he had memorized the entire index of a set of encyclopedias."[29]

However, Dr. Treffert's studies of Peek's brain also revealed major deficiencies and deformities: "Most striking was the total absence of the corpus callosum, the large connecting structure between the left and right hemispheres of the brain. Other connecting structures such as the left and right anterior and posterior commissures were missing as well. And there was extensive damage to the cerebellum, particularly on the right side."[30]

How was Peek able to perform in superhuman ways with such impairments to his brain? Materialism hasn't been able to explain it.

Peek isn't the only savant with unexplainable abilities, however. Dr. Treffert reports that there are fewer than 100 living savants with truly prodigious abilities.[31] For example, neurologist Oliver Sacks reported on autistic savants who couldn't perform primary math but were able to recite prime numbers with up to ten digits. As Dr. Ed Kelly recalls it: "Sacks was able to verify the primacy up to ten digits, but only by means of published tables, while the twins themselves went on happily exchanging numbers of progressively greater length, eventually reaching twenty digits. Sacks makes the intriguing suggestion that they cannot literally be *calculating* these enormous numbers but may instead be *discovering* them by navigating through some vast inner imaginal landscape in which the relevant numerical relations are somehow represented pictorially"[32] [emphasis in original].

Dr. Powell recalls a child she worked with, who at age two could read and speak eight different languages.[33] Similarly, Dr. Larry Dossey recalls one of Dr. Treffert's patients: "A savant whose conversational vocabulary was limited to some 58 words but who could accurately give the population of every city and town in the United States with more than 5,000 people; the names, number of rooms, and locations of 2,000 leading hotels in America; the distance from any city or town to the largest city in its state; statistics concerning 3,000 mountains and rivers; and the dates and essential facts of more than 2,000 leading inventions and discoveries."[34]

In some cases, individuals are born with savant syndrome, and in other cases it is acquired. For example, Dr. Treffert describes a "middle-aged woman who had a stroke from which she fully recovered, except now she speaks with an unmistakable and precise foreign accent of a country she has never visited." In another case, a 54-year-old surgeon was struck by lightning and thereafter developed an "obsessive interest in classical music which was not present pre-incident" and now plays piano professionally.[35]

Savants pose perplexing questions for anyone studying the connection between the brain and consciousness. It is not understood how individuals with such brain impairment can simultaneously possess such remarkable mental abilities. As stated by Dr. Treffert: "No model of brain function, including memory, will be complete until it can fully incorporate and explain this jarring contradiction of extraordinary ability and sometimes permeating disability in the same person. Until we can fully explain the savant, we cannot fully explain ourselves nor comprehend our full capacities."[36]

Dr. Treffert isn't sure how to explain this phenomenon even though he's been studying it since 1962. He comments: "I'm not entirely closed to the possibility that there is some link to a universal knowledge.…It seems like savants, especially autistic savants, know so many things they haven't learned, and it's almost as if they tapped into a universal knowledge."[37]

In other words, he's alluding to nonlocal consciousness. And as we will see in chapter 5, there is emerging evidence that certain savants also possess strong telepathic abilities.

Together, the findings make us wonder: Are savants' brains structured in a way that allows them to access information from outside their bodies?

Shuffling animal brains

Studies on animal brains raise additional questions about the connection between the brain and consciousness. American neuropsychologist Karl Lashley, PhD's work from the 1920s is instructive. Dr. Lashley's aim was to understand where in the brain memories are stored. He trained rats to perform certain tasks, such as jumping "through miniature doors to reach a reward of food."[38] He then systematically destroyed their brains, one part of the brain at a time. Miraculously, the rats could still perform the tasks.

As author Lynne McTaggart reports: "Their motor skills might be impaired, and they might stagger disjointedly along, *but the rats always remembered the routine*"[39] [emphasis in original]. Dr. Lashley was unable to show where in the brain memories were stored because he destroyed everything and yet the rats could still remember the routine. How could that be? Maybe he unwittingly uncovered that the brain-consciousness connection needs to be more closely examined. Summing it up, he said, "If I didn't know better, I would think that memory is stored outside of the brain."[40]

Then came psychologist Karl Pribram, PhD, who worked with Lashley at Yale University and subsequently left for Stanford. He was intrigued by Dr. Lashley's findings and advanced the studies. He found that monkeys with damage to their brains' frontal cortexes were able to perform tasks as successfully as monkeys with no damage.[41] Like Dr. Lashley, Dr. Pribram was perplexed by the inability to isolate the relationship between brain activity and consciousness.

University of Indiana anatomy professor Paul Pietsch was skeptical when he heard Lashley's and Pribram's claims. Pietsch referred to himself as a "materialist," saying "Lashley's principles seemed like a coverup" and that Lashley must have simply "concocted his doctrines."[42] Pietsch sought to disprove Lashley's and Pribram's theories by damaging salamander brains and examining whether they still exhibited feeding behavior.[43] To his surprise, no matter what he did to the salamanders' brains, they not only lived but also exhibited feeding behavior. McTaggart summarizes the results: "In more than 700 experiments, Pietsch cut out scores of salamander brains. Before putting them back in, he began tampering with them. In successive experiments he reversed, cut out, sliced away, shuffled and even sausage-ground his test subjects' brains. But no matter how brutally mangled, or diminished in size, whenever, whatever was left of the brains were returned to his subjects and the salamanders had recovered, they returned to normal behavior."[44]

As Pietsch summarizes in his 1981 book, *Shufflebrain*: "The animals invariably fed the moment they recovered from postoperative stupor, no matter which region I removed. Massive destruction of the brain reduced feeding but did not stop it."[45] Pietsch even found that when he replaced part of the salamander's brain with a tadpole's brain, it still exhibited feeding behavior.[46] In 1973, Pietsch was featured on the television show *60 Minutes*, where he discussed his findings.

Pietsch laments that he had to give up his staunchly held materialist stance: "I had complete faith, too, that my science would one day write the most important scientific story of all: how a brain gives existence to a mind. But I was wrong. And my very own research, which I call *shufflebrain*, forced me to junk the axioms of my youth and begin my intellectual life all over again"[47] [emphasis in original].

Memory transmitted via nonbrain organs

Now let's take the brain out of the picture entirely.

In a number of organ-transplant cases, recipients of the organs inherited memories and personality characteristics of the original donor. And in all reported cases, the organ donated was *not* the brain.

Psychoneuroimmunologist Dr. Paul Pearsall explored such cases. In one case, an eight-year-old girl received the heart of a ten-year-old girl who had been murdered. Neither the heart recipient nor her family knew that the donor had been murdered. Shortly after her transplant procedure, the eight-year-old girl began having persistent nightmares; she would wake up in the middle of the night with vivid details of a murder. The nightmares were so severe that the girl's mother took her to a psychiatrist, to whom the girl described the murder in detail (time, weapon, clothes). The psychiatrist contacted the police, and they were able to find the murderer and arrest him. The girl's information was proven to be fully accurate.[48]

In another case, Dr. Pearsall interviewed a 52-year-old man who received the transplanted heart of a 17-year-old boy. After the transplant, the man said, "I could never understand it. I loved quiet classical music before my new heart. Now, I put on ear phones, crank up the stereo, and play loud rock-and-roll music. I love my wife, but I keep fantasizing about teenage girls. My daughter says I have regressed since my heart surgery, and that I act like a sixteen-year-old." His daughter comments: "It is really embarrassing sometimes. When my friends come over, they ask if my dad is going through his second childhood. He's addicted to loud music, and my mom says the little boy in him is finally out."[49]

Science does not understand how this can happen. Cardiologist Pim van Lommel, MD, laments: "Unfortunately, the reservations of transplant centers and transplant organizations have so far prevented any systematic scientific research into this now-and-then reported phenomenon."[50]

In the absence of understanding *how* these phenomena occur, they undoubtedly pose a threat to materialism because memories, preferences, and behaviors were transmitted without involving the brain.

Chapter Summary

❍ The definition of consciousness depends on who you ask. For the purposes of this book, consciousness refers to the mind, one's sense of awareness, and inner experience; it is the sense of experiencing life.

❍ A foundational question explored in this book is: "Does the brain produce the consciousness?" Mainstream materialist science assumes that the brain produces consciousness, but doesn't know how this occurs.

❍ However, some scientists view the brain not as the producer of consciousness but more like an antenna or a filter of consciousness. In other words, consciousness exists independently of the body.

❍ If the brain produces consciousness, we might expect heightened conscious experience to correlate with increased brain activity. However, that is not what we find:

— A 2012 study suggests that psychedelic mushrooms reduced brain activity, even though the experiences are enriched and hyper-real. Similarly, people report having lucid, hyper-real experiences with little or no brain function (near-death experiences).

— In cases of terminal lucidity, patients with brain disorders exhibit sudden and unexplained clarity near the time of death.

— Savants who are disabled in some ways have superhuman memories and abilities in math, music, and other areas. Some savants might even be highly telepathic.

— Systematically destroying and replacing elementary animal brains does not seem to eliminate memory and certain behaviors.

  - Organ-transplant patients sometimes take on the memories and personalities of their donors, and in these cases the organ they've received is not the brain.

- Collectively, the evidence suggests that we may need to rethink the materialist claim that the brain produces consciousness.

Chapter 3

Quantum, Relativistic Chaos
Proven and Accepted Science that Defies Common Sense

Quantum theory does not allow a completely objective description of nature.[1]

—Werner Heisenberg, Nobel Prize-winning physicist

At the quantum level, reality does not exist if you are not looking at it.[2]

—Andrew Truscott, associate professor at the Australian National University

The interpretation [of quantum mechanics] has remained a source of conflict from its inception. For many thoughtful physicists, it has remained a kind of "skeleton in the closet."[3]

—J. M. Jauch, physicist

To describe what has happened, one has to cross out that old word "observer" and put in its place the new word "participator." In some strange sense the universe is a participatory universe.[4]

—John Wheeler, physicist

But obviously, our present actions cannot determine the past. *The past is the "unchangeable truth of history." Or is it?*[5] [emphasis in original].

—Physicists Bruce Rosenblum and Fred Kuttner

The difficulty really is psychological and exists in the perpetual torment that results from your saying to yourself, "But how can it be like that?" which is a reflection of uncontrolled but utterly vain desire to see it in terms of something familiar....Do not keep saying to yourself..."But how can it be like that" because you will get...into a blind alley from which nobody has yet escaped. Nobody knows how it can be like that.[6]

—Richard Feynman, Nobel Prize-winning physicist describing the dissonance between quantum physics and everyday human experience

Universal nonlocality offers us profound evidence that our Universe is fundamentally interconnected as a unified entity.[7]

—Jude Currivan, cosmologist

In the previous chapter, we established that there is a true question as to the relationship between the brain and consciousness. Maybe we need to rethink materialism. Hold that thought.

We will take a detour for one chapter into the strangeness of proven science before we dive into concepts related to psychic abilities and survival of bodily death. This science is the underpinning of our reality and needs to be considered first. My aim is to simplify the topics as much as possible because they can be confusing.

The pros and cons of following common sense

If there is anything to take away from this chapter, it is that our perceptions can lead us astray because we live in a reality far more mysterious than our everyday senses show us. Just because something doesn't seem to "make sense" doesn't mean it is not real. We can use our everyday perceptions for an *approximation* of reality, but not a 100 percent accurate picture. Physicists Dr. Stephen Hawking and Dr. Leonard Mlodinow explain: "Common sense is based upon everyday experience, not upon the universe as it is revealed through the marvels of technologies such as those that allow us to gaze deep into the atom or back to the early universe."[8]

For example, when you talk on your cell phone, it wirelessly connects with another person's device that is physically distant from yours. You can't see a connection between the phones with your eyes. Yet somehow the phones communicate. If you relied solely on your everyday senses, you might not believe that cell phones could communicate wirelessly via radio waves that are invisible to the naked eye. But we now know from science that our eyes only see a tiny fraction of light waves on the electromagnetic spectrum. There are many types of light that our eyes don't see. But they exist, regardless (see the figure below).

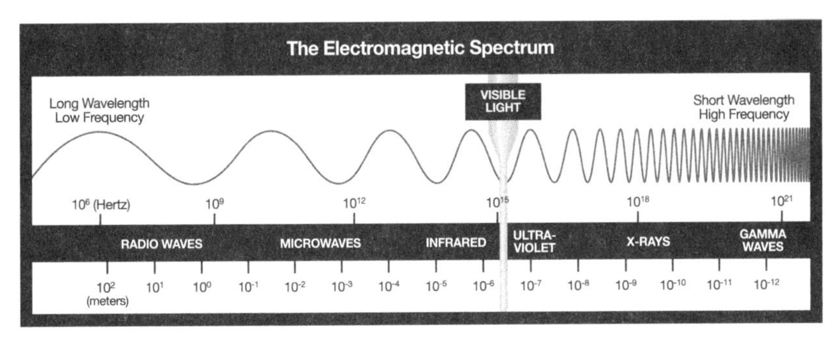

Visible light is only a tiny fraction of the electromagnetic spectrum. Most of the spectrum exists beyond what our ordinary vision can detect. Our eyes show us only a sliver of reality.[9]

As stated by Drs. Kelly and Presti: "We respond visually to just a tiny segment of the electromagnetic spectrum, and we hear a range of acoustic frequencies that can only be described as impoverished relative to the hearing of our dogs and cats. Our chemosensory capacities, taste and smell, are even more radically impoverished relative to those of many other creatures. In effect, we have been adapted by biological evolution in such fashion that under everyday circumstances we inhabit an experienced world that *represents but a tiny fraction of what actually exists*"[10] [emphasis added].

Making things more difficult, our brain can trick us. Our brains often perceive the world *in*accurately. Donald Hoffman, professor of cognitive science at the University of California, Irvine, shares his findings of evolutionary game theory:

> If you have an organism that sees reality as it is and is competing with an organism that sees none of reality but is only tuned to the fitness consequences in its environment... *then the organism that sees reality as it is can never win....* Evolution does not shape perceptual systems to see reality as it is. It shapes our perceptual systems simply to keep us alive

> long enough to have kids....So that means that I can't trust that my perceptions of space and time and physical objects are an insight into the nature of reality as it is. Rather, what evolution is telling us...is that our perceptual systems are a species-specific adaptation not designed to show us reality as it is, but in fact shaped to hide reality, because we don't need to know reality; it's unnecessary[11] [emphasis added].

So without assistance from science and technology, we are not in the best position to opine on what "reality" is just by following our programming.

Quantum vs. Newtonian worldviews

With that preface, let's examine an area of science that defies our everyday senses, but which has been proven to be real: "quantum mechanics" (alternatively called "quantum physics"). Quantum mechanics was conceived in the early 1900s. It deals with how tiny particles—pieces of physical matter—behave. They don't behave in ways that we would expect based on our everyday experiences. But since large objects, including our bodies, are made from lots and lots of small particles, it is important to study how small particles behave.

Quantum mechanics has proven to be a highly successfully theory. Physicists Bruce Rosenblum and Fred Kuttner remind us in their book *Quantum Enigma*: "Quantum theory works perfectly; no prediction of the theory has ever been shown in error. It is the theory basic to all physics, and thus to all science. One-third of our economy depends on products developed with it. For all *practical* purposes, we can be completely satisfied with it. But if you take quantum theory seriously *beyond* practical purposes, it has baffling implications"[12] [emphasis in original]. Further, it "is the most battle-tested theory in all of science. It has no competitors.... Niels Bohr, a founder of quantum theory, warned that unless you're shocked by quantum mechanics, you have not understood it."[13]

Physicist Brian Greene sums it up well: "Common experience—mundane, ordinary, day-to-day activities" are "part of a classical charade, hiding the true nature of our quantum world." Our everyday world is "nothing but an inverted magic act, lulling its audience into believing in the usual, familiar conceptions of space and time, while the astonishing truth of quantum reality lay carefully guarded by nature's sleights of hand."[14]

The "classical charade" is a reference to "Newtonian" physics, the branch of physics that came from Sir Isaac Newton (1642–1727). It deals with large, visible objects that behave more in-line with everyday perceptions.

Classical, Newtonian physics does a good job of approximating the world around us. As physicists Dr. Stephen Hawking and Dr. Leonard Mlodinow put it: "Classical theories such as Newton's are built upon a framework reflecting everyday experience, in which material objects have an individual existence, can be located at definite locations, follow definite paths, and so on."[15] But as we'll see, the quantum world doesn't act in a classical, Newtonian manner.

Most nonphysicists are trained under a set of Newtonian assumptions, in spite of its limitations. The approximations are usually good enough. But if Newtonian physics is imperfect, what might nonphysicists be missing by glossing over the quantum reality? As stated by former Cornell psychologist Dr. Daryl Bem: "The phenomena of modern quantum physics are…mind-boggling, but they are so technical that most non-physicists don't know about them."[16]

And further elaborated upon by Dr. Kelly: "Few working psychologists and neuroscientists, let alone the public at large, have any conception of the fundamental significance of quantum theory. Classical concepts and approximations are often sufficient to support the concerns of the special physical sciences, and quantum mechanics is scarcely mentioned in the context of general education even at a college level. Yet it cannot be emphasized too strongly that the classical physics consensus that underwrites practically everything now going on in psychology, neuroscience, and philosophy of mind has in fact been completely undermined by this tectonic shift in the foundations of physics."[17]

Now let's get into the quantum phenomena themselves. I've stripped away the technical details so you can understand the main points.

Entanglement—"spooky action at a distance"

One of the primary tenets of quantum mechanics is known as "entanglement": the finding that the states of two *distant* particles mirror each other *instantaneously*. So entangled particles are in some way connected, even if they are physically separated—no matter how far apart they are in space and time.[18]

How can this be? The particles aren't next to each other, and yet there is an instantaneous effect. Albert Einstein postulated that nothing can travel faster than the speed of light. The speed of light is nearly 300 million meters per second. But even that is not as fast as "instantaneous." For this reason, Einstein didn't like entanglement, calling it "spooky action at a distance." He was skeptical, stating: "Quantum mechanics is certainly imposing. But an inner voice tells me that it is not yet the real thing."[19]

It's then perhaps not surprising that Einstein tried to disprove quantum entanglement. In 1935 he conducted the famous Einstein-Podolsky-Rosen experiment (known as the EPR paradox).[20] Ironically, not only was he unable to disprove entanglement, his experiment only further proved its reality.[21] Since then, evidence has continued to mount. For example, in 2015 *The New York Times* profiled one such study in an article entitled "Sorry, Einstein. Quantum Study Suggests 'Spooky Action' Is Real."[22]

The findings suggest there is a hidden, fundamental interconnectedness—or "nonlocality"—in the universe that our eyes don't see. Put another way by physicists Rosenblum and Kuttner: "That our actual world does not have separability is now generally accepted, though admitted to be a mystery. In principle, any objects that have ever interacted are forever entangled, and therefore what happens to one influences the other. Experiments have now demonstrated such influences extending over more than one hundred kilometers. Quantum theory has this connectedness extending over the entire universe."[23]

"Spooky," indeed.

Physicist David Bohm went a step further and suggested an interconnectivity of everything: "One is led to a new notion of *unbroken wholeness* which denies the classical idea of analyzability of the world into separately and independently existing parts.... We have reversed the usual classical notion that the independent 'elementary parts' of the world are the fundamental reality, and that the various systems are merely particular contingent forms and arrangements of these parts. Rather, *we say that inseparable quantum interconnectedness of the whole universe is the fundamental reality*, and that relatively independently behaving parts are merely particular and contingent forms within this whole"[24] [emphasis added].

How might this "spooky," nonlocal entanglement relate to consciousness? Dr. Dean Radin, the chief scientist at the Institute of Noetic Sciences, suggests in his 2006 book, *Entangled Minds*, "that we take seriously the possibility that our minds are physically entangled with the universe." He concedes: "I'm not claiming that quantum entanglement magically explains all things spooky. Rather I propose that the fabric of reality is comprised of 'entangled threads.'"[25]

This notion is one worth considering in the context of the "nonlocal consciousness" phenomena that we will explore.

The observer affects reality

The famous double-slit laser experiment, first carried out in 1927 by Clinton Davisson and Lester Germer at Bell Labs, illustrates another spooky quantum phenomenon.[26] Physics Nobel Prize winner Richard Feynman has said the experiment is "impossible, *absolutely* impossible to explain in any classical way" [emphasis in original], and it "has in it the heart of quantum mechanics." He further states, "We cannot make the mystery go away."[27]

This experiment is difficult to describe in writing, and it is even more difficult to comprehend because it makes no sense. So in the spirit of keeping things simple, I will skip the details and stick with a basic description.

The first strange finding is that *particles* (i.e., "compact lumps"[28]—pieces of matter that have a finite location) can act like *waves* of probability (i.e., they have a distributed, uncertain location; maybe here, maybe there). My everyday experience tells me that solid particles are always solid particles. How can it act like a particle sometimes and a wave at other times? When I look at my table, it appears to have a finite location. It is by my kitchen. It isn't "sometimes in my living room," "sometimes in my bedroom." So how is it that particles of matter act like waves?

It gets even stranger.

Whether the particle behaves like a particle or a wave depends on whether it is observed. When the experimenter "looks" at the double slit in the experiment, particles act like particles. When the observer doesn't look, particles behave like waves.

This is known as "the collapse of the wave function"—when an observer "looks," the wave "collapses," and the observer sees a particle. For some unknown reason, the simple act of looking changes a particle's behavior.

Rosenblum and Kuttner explain: "If someone looked in a particular spot and happened to see the atom [i.e., particle] there, that look 'collapsed' the spread-out waviness of that atom to be wholly at a particular spot.... Nevertheless, the waviness of that atom existed at that different spot immediately before the...observer collapsed it." In other words, the particle behaves like a wave until it's observed, at which point it behaves like a particle. Rosenblum and Kuttner ask rhetorically, as you might also be asking yourself, "Observing an atom being in a particular place *created* its being there? Yes."[29]

Physicist Brian Greene similarly acknowledges the strangeness of this effect: "I understand full well if this...leaves you shaking your head.... Imagine Lucille claiming she's a blonde—until someone looks, when she immediately transforms into a redhead."[30]

The mere act of observing changes the behavior of physical matter. Particles behave like particles when they are observed and like waves when they are not observed. As Physicist John Wheeler summarized it: "No phenomenon is a real phenomenon until it is an observed phenomenon."[31] Put another way by Wheeler: "Nothing...exists until it is observed."[32]

But what do physicists mean by "observing"? It seems like a vague term. Does someone need to look with one's eyes in order to convert wave-like behavior into particle-like behavior? Or does the wave function collapse occur because the observer uses an electronic measuring device, and that device somehow impacts the experiment?

If the first explanation—simply "looking" with one's eyes—is what causes the collapse of the wave function, then perhaps consciousness is involved. The act of looking implies that the conscious mind enters the picture.

However, many physicists reject the idea that consciousness should enter the picture at all. For example, physicist Dr. Stephen Hawking stated: "I get uneasy when people, especially theoretical physicists, talk about consciousness."[33] Physicist Neil deGrasse Tyson comments: "I wonder whether there really is no such thing as consciousness at all."[34] Consciousness is a topic typically not widely emphasized in the realm of physics. It's normally explored by neuroscientists, psychologists, and philosophers. Physicists aren't trained to think about it. It's outside their domain of expertise.

For this reason, mathematician and astronomer Bernard Carr states: "While the *contents* of consciousness are certainly of interest to science, most physicists assume that the study of consciousness itself is beyond their remit, physics being concerned with the 'third-person' rather than 'first-person' account of the world"[35] [emphasis in original].

But some prominent physicists haven't been so quick to dismiss consciousness as the potential culprit in the collapse of the wave function. Nobel Prize-winning physicist Eugene Wigner said, "The being with a consciousness must have a different role in quantum mechanics than the inanimate measuring device."[36] He also commented, "It is the entering of an impression into our consciousness which alters the wave function,"[37] adding, "it is at this point that the consciousness enters the theory unavoidably and unalterably."[38]

Quantum physicist Amit Goswami agrees: "We cannot understand quantum physics without introducing consciousness into it."[39] Other prominent physicists such as John von Neumann and Henry Stapp have taken a similar stance.[40] Dr. Stephen Hawking's former coauthor, Roger Penrose, is even talking about consciousness. Penrose, with his co-collaborator, anesthesiologist Dr. Stuart Hameroff, declared: "We conclude that consciousness plays an intrinsic role in the universe."[41]

And perhaps most significantly, Max Planck, one of the founders of quantum physics, stated in 1931: "I regard consciousness as fundamental. I regard matter as derivative from consciousness. We cannot get behind consciousness. Everything we talk about, everything that we regard as existing, postulates consciousness."[42]

The debate continues. Science isn't sure if consciousness plays a role in affecting a particle's wavelike behavior. Physicist Lucien Hardy proposed a study that could directly test this idea, as described in a May 2017 *New Scientist* article. The article notes how significant the results would be: "If such an experiment showed deviations from quantum mechanics, it could provide the first hints that our minds are potentially immaterial."

Swiss physicist Nicolas Gisin agrees: "If someone does the experiment and gets a surprising result, the reward is enormous. It would be the first time we as scientists can put our hands on this mind-body or problem of consciousness."[43]

The design would ask participants to "put their minds" to the double slit. If their minds alone could affect the wavelike behavior, then perhaps consciousness *is* involved. One would think that an experiment like this would fail. How could mental attention do anything outside of the mind? Using this line of thinking, Gisin commented that for a study like this: "There is an enormous probability that nothing special will happen, and that quantum physics will not change."[44]

What Hardy and Gisin don't realize is that the test they proposed has already been conducted. And contrary to Gisin's above statement, the results so far show that something special *does* happen.

Studies on this topic were conducted recently at the Institute of Noetic Sciences (IONS) in Petaluma, California, and were led by chief scientist Dr. Dean Radin. He ran 17 studies over an eight-year period. Apparently, as evidenced by the *New Scientist* article, the word hasn't gotten out to the mainstream physics community.

Now to the details of the IONS studies. In early iterations, Dr. Radin asked meditators simply to focus their mental attention on the experiment. Guess what? They impacted the wave function, apparently with their minds alone.

Then Dr. Radin asked people far away from the site of the experiment to do the same thing. The double-slit laser equipment was in a lab in California, but participants in the study were all over the world. Somehow, just by mentally focusing on the experiment, participants impacted the wave function in the lab in California—no matter how far away they were. Interestingly, Dr. Radin also tested whether a Linux-based computer would be able to impact the wave function. It did not produce an effect (what does this say about whether artificial intelligence machines have consciousness?).[45]

Dr. Radin's staggering results have been published in two peer-reviewed science journals: *Physics Essays* (2012 and 2013)[46] and *Quantum Biosciences* (2015).[47] His results were not just marginally statistically significant, either. They were extremely significant. Depending on the way the data was combined, he got between four- and eight-sigma results across the studies (i.e., very strong). The statistics were confirmed by two independent evaluators, one of whom published his results in *Physics Essays*. At a 2016 lecture, Dr. Radin noted that the European Organization for Nuclear Research (CERN) won the Nobel Prize for its groundbreaking finding of the Higgs particle. CERN got five-sigma statistical results. Dr. Radin jokingly remarked, "We got a five-sigma result too, but I haven't heard from the Nobel Prize committee yet." But as a consolation, Dr. Radin did win the Nascent Systems Inc 2015 Research Prize. When Dr. Radin tweeted about his award on Twitter (December 2015), the tweet only received 17 "likes"![48] These groundbreaking studies haven't received the attention they deserve.

Gabriel Guerrer of the University of São Paulo is the first scientist to attempt to independently replicate the results. As the results were coming in, he commented to Dr. Radin, "In the last days it has been an intense mixture of feelings. I'm oscillating between OH MY GOD and wait, something must be wrong."[49]

Guerrer summarizes his findings in a March 2018 paper: "A post hoc combination of the formal experiments' scores...provided statistically significant results favoring the existence of anomalous interactions between conscious agents and a physical system. Further studies are warranted to formally test the post hoc hypothesis."[50] Wow.

If Dr. Radin's results can be replicated consistently, he probably does deserve a Nobel Prize. He would be moving science toward an answer to one of its greatest mysteries. The findings would validate the statements made by Planck, Wigner, and others, while turning most of science's world upside down.

An inference one might make from these findings is that consciousness is somehow *creating* particles of matter from waves of probability. That idea would fit with the framework discussed in the preface: Matter is a product of (and influenced by) consciousness. Further experimentation is needed to shed light on this theory.

Matter isn't what it seems

The double-slit study additionally raises questions about what "matter" really is. Our everyday human experience tells us that matter is easy to define: It is simply solid material that makes up the universe. My arm, which is made of many atoms of matter, certainly feels "solid" to me.

But science tells a different story. Science shows that atoms are 99.99999999 percent *empty* space. As stated by Dr. Diane Powell: "The nucleus of an atom takes up as much room in the atom as an ant on a football field."[51] Or as British physicist Sir Arthur Eddington said: "Matter is mostly ghostly empty space."[52]

If I simply relied on my everyday experience, I would have said my arm is solid and not mostly empty. But it *is* mostly empty. And the remaining 0.000000001 percent that isn't empty isn't exactly solid, either. As we saw in the double-slit study, sometimes particles behave like bits of matter, and at other times they behave like waves of probability. So this stuff that we call "matter" and assume to be solid, isn't actually solid at all.

As stated by physicist Peter Russell: "With the development of quantum theory, physicists have found that even subatomic particles are far from solid. In fact, they are nothing like matter as we know it. They cannot be pinned down and measured precisely. Much of the time they seem more like waves than particles. They are like fuzzy clouds of potential existence, with no definite location. Whatever matter is, it has little, if any, substance."[53]

For this reason, German physicist Hans-Peter Dürr stated: "Matter is not made of matter."[54] Or as physicist Fritjof Capra puts it: "Atoms consist of particles and these particles are not made of any material stuff."[55]

Furthermore, the Heisenberg Uncertainty principle (1927), conceived by Nobel Prize winner Werner Heisenberg, shows that we can't simultaneously know a particle's location and its momentum. If we measure its location, we can't know its exact momentum; if we measure its momentum, we cannot know its exact location.[56] So we are inherently limited in what we can know about matter. Capra summarizes it well: "The important point now is that this limitation has nothing to do with the imperfection of our measuring techniques. It is a principle limitation which is inherent in the atomic reality."[57]

These concepts do not jibe with Newtonian approximations of reality. Newtonian physics posits a simplistic view that aligns with 18th-century philosopher David Hume's "billiard ball" model of reality. That model implies that the interactions of matter are just like those of billiard balls. The billiard balls have finite locations, and we can measure the properties when the balls interact. The model implies, "That if we knew all of the equations governing the spatial positions of fundamental particles as a function of time, along with their initial conditions, then we would know everything that there is to know about reality."[58]

But as we've seen, this simplistic billiard-ball picture doesn't hold anymore. The quantum reality is far stranger than what we had for so long assumed. As summarized by Capra: "In quantum theory we have come to recognize probability as a fundamental feature of the atomic reality which governs all processes, and even the existence of matter. Subatomic particles do not exist with certainty at definite places, but rather show 'tendencies to exist,' and atomic events do not occur with certainty at definite times and in definite ways, but rather show 'tendencies to occur.'"[59]

Time isn't what it seems

One of Albert Einstein's most well-known theories has to do with "relativity," which suggests that time—how fast or slow one's "clock" goes—is not fixed. Instead, it is relative.[60] Time moves more quickly or slowly, depending on: a) how fast someone is moving; and b) how strong the experienced gravitational force is. The effect is known as "time dilation."

Why don't we experience "relative time" on a daily basis? It seems like time is moving at the same rate for me as it is for everyone else. My clock acts the same as my neighbor's clock. Normally we don't experience the effects because in our daily lives we're all traveling at roughly the same speed and are experiencing roughly the same gravitational force on Earth. However, the effect *does* exist in our daily lives; it's just so small that our ordinary senses don't notice it. For instance, an experiment in 1971 demonstrated that a clock on a flying airplane (i.e., traveling at high speeds) moved more slowly than clocks on the earth's surface—but only by roughly 180-billionths of a second.[61] So if you're a frequent flier, technically you're aging more slowly than the rest of us (but not by much).

There are cases in which the effects are more dramatic. One example is in the science-fiction film *Interstellar*. Matthew McConaughey's character travels to a planet with a much higher gravitational force than Earth's, so he experiences time more slowly than those on our planet. He travels back to Earth not having aged much, but finds that his previously young daughter is now an old woman on her deathbed.

But at least "time dilation" preserves the direction of the arrow of time. Time goes from past to present to future, even if clock speeds differ. Quantum physics makes us question even that basic assumption. Is it true that time only moves from past to present to future?

Maybe not.

In "delayed choice" thought experiments originally conceived by John Wheeler, the choice of the present theoretically *changes the past*. A particle's behavior is impacted when the experimenters make a decision about which experiment to run, *after* a particle had already taken a particular path. The future decision impacts the *past* behavior of the particles.[62]

In 2015, physicists at the National Australian University seem to have confirmed this effect empirically; they claim that the future measurement impacts an atom's past. As stated by the researchers: "The atoms did not travel from A to B. It was only when they were measured at the end of the journey that their wave-like or particle-like behavior was brought into existence."[63] As Wheeler summarizes: "We have a strange inversion of the normal order of time."[64]

While additional confirmation will be needed, the potential implications are profound. They suggest that the relationship between past, present, and future is less clear than our ordinary experience teaches us. And as we will discover later in this book, consciousness sometimes displays similar oddities with regard to time.

With this new understanding of time, consider transpersonal psychologist Dr. Carl Buchheit's take: "The present is the future's past, and is therefore what has already occurred on the way to a (now) pre-existing future fulfillment....It is the future that is doing the creating of a past congruent with itself and it is the present that chooses itself to be congruently the past of a particular future."[65]

And even Einstein said: "People like us, who believe in physics, know that the distinction between past and present and future is only a stubbornly persistent illusion."[66]

Space isn't what it seems

We also learn from Einstein's relativity theory that *space* is not fixed. This notion is completely counterintuitive. My table is eight feet long. I would expect that it would be eight feet long in all circumstances (unless someone were to break it). But in relativity theory, length shrinks. This effect is known as "length contraction."

Imagine that my eight-foot-long table is on a spaceship that is moving close to the speed of light. From the perspective of a stationary observer, the table's length will appear to be less than eight feet.[67] We don't normally notice this effect because in our daily lives, no one is traveling near the speed of light.

Relativity thus teaches us that both time and space are relative. They are not fixed. The Newtonian conception of a "universe with a rigid, unchangeable arena" is not how things work. As Brian Greene reminds us: "According to Newton, space and time supplied an invisible scaffolding that gave the universe shape and structure."[68] Not so in the quantum, relativistic reality.

Chaos theory

It's not just quantum physics and relativity theory that show flaws in our everyday perceptions of reality. We also need to look at chaos theory. As stated by James Gleik in his 1987 book, *Chaos: Making a New Science*: "Where chaos begins, classical science stops,"[69] and "Chaos cuts away at the tenets of Newtonian physics."[70]

Chaos theory was uncovered by meteorologist Edward Lorenz in 1961. While predicting weather patterns, he rounded one of the numbers in his mathematical equations from 0.506127 to 0.506. No big deal, right? The numbers are close enough. However, simply by rounding the number by a tiny amount, the weather predictions his equations yielded were wildly different.[71] What Lorenz uncovered, through the use of advanced computing, was that tiny changes in initial conditions can have a huge effect on eventual outcomes. This idea is known as the "Butterfly Effect": "the notion that a butterfly stirring the air today in Peking can transform storm systems next month in New York."[72] Mathematically, that's how dramatic minor changes in initial conditions can be.

The Butterfly Effect is a metaphor that we can use to illustrate one of the key tenets of chaos theory: The world operates in a *nonlinear* fashion. If the world functioned linearly, the seemingly minor effect of a butterfly flapping its wings in China, moving the air particles around ever so slightly, would yield a proportionally small effect on the weather. That's not how the math works though. Rather, the butterfly's movement has a *disproportional, nonlinear* effect on the weather. As doctor and philosopher Sir David Hawkins, MD, PhD, summed it up: "All life processes are, in fact, nonlinear."[73]

If we followed our everyday perceptions, we would think the world is *linear, Newtonian, and fixed*. But real, proven science instead suggests that our reality is *nonlinear, quantum, and relativistic*.

With this counterintuitive reality in mind, we are now ready to explore counterintuitive phenomena of consciousness.

Chapter Summary

- Following our ordinary senses can be dangerous: they give a good approximation of reality but miss the full picture.
- Classical, Newtonian physics provides useful approximations of reality.
- The fields of quantum physics and relativity, conceived in the early 1900s, don't align with everyday experience, but are accepted as being real.
  - Particles that are physically distant can have invisible connections (known as "entanglement").
  - An "observer" can change the behavior of a particle, simply by observing.
  - Matter isn't solid.
- Time moves at different rates depending on one's frame of reference. For example, at high speeds, clocks slow down ("time dilation"). Additionally, there is a question as to whether it only goes from past to present to future.
  - Length changes depending on one's frame of reference. For example, at high speeds, length shrinks ("length contraction"). Like time, space is not a fixed quantity.
- Chaos theory teaches us that tiny differences in initial conditions can have a massive impact on final outcomes—a concept known as nonlinear dynamics.
- Contrary to what our ordinary perceptions teach us, the universe is nonlinear, quantum, and relativistic rather than linear, Newtonian, and fixed.

Section III

Wizard-like Abilities? Scientific Evidence

This section discusses independent areas of research on psychic phenomena. The accumulated evidence suggests that humans (and animals) have psychic abilities, even if they are sometimes subtle.

CHAPTER 4

Remote Viewing
Sensing from a Distant Location

I never liked to get into debates with the skeptics, because if you didn't believe that remote viewing was real, you hadn't done your homework.

We didn't know how to explain it, but we weren't so much interested in explaining it as in determining whether there was any practical use to it.[1]

—Major General Edmund R. Thompson,
U.S. Army assistant chief of staff for Intelligence, 1977–1981;
deputy director for Management and Operations, DIA, 1982–1984

While working for the CIA program at our lab in Menlo Park, California, our psychic viewers were able to find a downed Russian bomber in Africa, to describe the health of American hostages in Iran, and to locate a kidnapped American general in Italy. We also described Soviet weapons factories in Siberia and a Chinese atomic-bomb test three days before it occurred and performed countless other amazing tasks—all using the ability that our colleague Ingo Swann dubbed remote viewing.[2]

—Russell Targ, laser physicist, cofounder of the Stanford Research Institute's $20 million program to explore psychic abilities in the 1970s and 1980s; holds two NASA awards for inventions in lasers

You can't be involved in this for any length of time and not be convinced there's something here.[3]

—Norm J, a former CIA official who tasked the
U.S. military's Fort Meade remote viewing unit

There were times when they wanted to push buttons and drop bombs on the basis of our information.[4]

—Hal Puthoff, a former manager of the U.S. government's remote viewing program

All I can say is that if the results were faked, our security system doesn't work. What these persons "saw" was confirmed by aerial photography. There's no way it could have been faked.…Some of the intelligence people I've talked to know that remote viewing works, although they still block further research on it, since they claim it is not yet as good as satellite photography. But it seems to me that it would be a hell of a cheap radar system. And if the Russians have it and we don't, we are in serious trouble.[5]

—Former U.S. Congressman Charles Rose, chairman of the House Sub-Committee on Intelligence Evaluation and Oversight (~1979)

She went into a trance, and while she was in a trance she gave some latitude and longitude figures. We focused our satellite cameras on that point and the [lost] plane was there.[6]

—Former U.S. president Jimmy Carter describing a remote viewer who helped the U.S. Air Force and CIA locate a lost Russian Tupolev-22 bomber that had crashed in a jungle in Africa

By the standards of any other area of science[,] remote viewing is proven.[7]

—Richard Wiseman, consulting editor for *Skeptical Inquirer* magazine and professor of the Public Understanding of Psychology at the University of Hertfordshire

Remote viewing (sometimes called "remote perception") refers to the ability to perceive something—with one's mind—from a distance.

Yes, that's right—remote viewing is the act of seeing or sensing something without physically being there to experience it. And what's more astonishing is that it can be done independent of time. Remote viewers are able to perceive at a distance and in the present, past, and future.

Dr. Claude Swanson, a physicist who holds a PhD from Princeton University, describes his amazement when he learned of remote viewing:

I came face [sic] with information that I could not dismiss and could not explain. I was riding in a car with a couple of scientists from Washington, D.C. They were talking about something I had never heard of before. It was called "remote viewing." They told me about a man who could go into a trance in a shielded room, and send his mind out into the world. They mentioned experiments where he could look into locked rooms, peer into locked safes, and read secret documents half way around the world…while never leaving his chair in the shielded room!

The scientists telling me this were people I knew and respected. This was not some rumor they were passing on. It was a project they had direct knowledge of. I could not dismiss this information, and yet it violated everything I had been taught about physics.

The experiments had been conducted in a heavily shielded room.…There was no known force which could explain these experiments.

Even more disturbing was the way it was done. The remote viewer did not simply receive a signal from someone outside. He described the experiment as though he actually left his body and *went* to the location. He could move around when he was there. He passed through walls of steel like they weren't there. In some cases, he described events a half hour *before* they happened. This was something new. It was a revolution in physics if it were true. It meant a new theory was needed.

I followed this up with further research to verify what I had heard. I discovered that there were published papers on this subject, in fact there was a great deal of confirming research, but I had never known to look for it before. Probably I wouldn't have believed it before. Once I knew that there was something important missing from the "old" physics, I began searching for more information. I tried to keep an open mind while at the same time remaining rigorously objective.

My education, as well as the influence of the media, caused me to be skeptical about all claims involving the paranormal, psychics, and other unconventional phenomena. The media

> had presented all these topics as being due to hoaxes and hallucinations. But as I investigated more deeply, I was amazed at how many top-quality researchers, M.D.s and Ph.Ds from top institutions, had performed research into psychic phenomena. And contrary to what I had always been taught, they found very real effects, solid experimental results that cried out for scientific explanation[8] [emphasis in original].

Under the mainstream, materialist assumption that "the brain produces consciousness," remote viewing seems paranormal and impossible. But as discussed in the preface, if consciousness is like a stream of water and an individual brain is a localized whirlpool, then having access to other parts of the stream (i.e., remote viewing) is possible.

Researchers often report that remote viewers go into a meditative trance while remote viewing. Taken in the context of chapter 2, this might make sense: If, in the trance, they are reducing brain activity through a calming of the mind, perhaps they are eliminating noise that ordinarily prevents access to the broader stream.

Stanford Research Institute (SRI)/Stargate Project (~1972–1995)

Remote viewing has been tested extensively, including a 24-year program sponsored by the U.S. government. The government program has been known by many names ("Grill Flame," "Center Lane," "Sun Streak"), but most recently it was called "Stargate." For years the details were kept classified, but over time information has been released.

Haven't heard of Stargate before? Maybe you just put this book down, Googled "Stargate," and saw the Wikipedia page stating that Stargate "was never useful in any intelligence operation. Information provided by the program was vague, included irrelevant and erroneous data, and there was reason to suspect that its project managers had changed the reports so they would fit background cues."[9]

An exploration of primary sources tells a different story.

The lead researchers were well-respected laser physicists Russell Targ and Hal Puthoff of the Stanford Research Institute (SRI). They knew of the purported remote viewing phenomenon and wanted to explore it further, but they needed funding. So they approached a contact at the

CIA, John McMahon, then Deputy Director for Intelligence, "who was well known for not suffering fools gladly."[10] McMahon entertained them because Targ and Puthoff were already known in the agency. McMahon said he would support their work if they could use remote viewers to provide useful information about a Soviet Union site. They succeeded. Remote viewing researcher Stephan A. Schwartz comments: "Just the fact that such a program came to exist at SRI is notable."[11] Targ and Puthoff were then given funding by the CIA, which they could use both for their own research and for government research, though the government's involvement was to be kept classified.

The program ran for many years and included studies demonstrating that remote viewing is in fact real. There are several striking examples worth noting. One includes a talented psychic named Ingo Swann. As Targ states in his 2012 book, *The Reality of ESP: A Physicist's Proof of Psychic Abilities*: "Why do I believe in ESP? Two of the main reasons come from my opportunities to sit with Ingo Swann in our laboratory in California. The first was when he drew pictures of a secret U.S. cryptographic site in Virginia, and the second was when he gave a stunning description of a Chinese atomic bomb test three days before it happened, with only the geographic coordinates for guidance."[12]

Targ and Puthoff studied Swann's abilities closely. Targ recalls what happened: "We put a laser in a box and asked Ingo to tell us whether it was on or off. We would ask him to describe pictures hidden in opaque envelopes or in a distant room. Ingo did all these tasks excellently, but he found them to be very boring. He told us many times that, if we didn't give him something more interesting to do, he was going back to New York and resume his life as a painter."[13]

So they gave him a more interesting task. They wanted to understand the physical limits of remote viewing, so they asked him to "see" the planet Jupiter from where he was sitting in California. They would then compare his description with what would be later shown by space probes circling Jupiter. At the time of the session on April 27, 1973, mankind did not yet know the specifics of what Jupiter looked like. Targ reminds us that, back then, "conventional scientific wisdom held that Jupiter did not possess any rings."[14] According to Targ's transcript of the official recording, Swann reported: "There's a planet with stripes.... Very high in the atmosphere there are crystals...they glitter. Maybe the stripes are like bands of crystals, maybe like rings of Saturn, though not far out like that. Very close within the atmosphere. I bet you they'll reflect radio probes."[15] Swann then drew what he was "seeing."[16]

Six years later (1979), results from the Voyager 1 probe were reported in *Time* magazine: "And most surprising of all, [Voyager 1] revealed the presence of a thin, flat ring around the great planet. Said University of Arizona astronomer Bradford Smith: 'We're standing here with our mouths open, reluctant to tear ourselves away.'"

Targ comments: "Unlike Saturn's rings, which are clearly visible from the Earth even through small telescopes, Jupiter's rings are very difficult to see. So difficult, in fact, that they weren't discovered until they were first confirmed by the Voyager 1 spacecraft in 1979."[17] He was able to see Jupiter "with his mind"...while in California?! How is that possible? Can anyone else do this?

Yes—there are others. For example, Joe McMoneagle worked with the government and helped to remotely locate a downed Russian Tupolev-22 bomber that U.S. satellites were unable to find. As Targ describes it:

> Joe was given a large map of Africa on which he could try to match and record his mental pictures as they emerged. The first thing he saw on his mental screen was a river flowing to the north. Working with his eyes alternately open and closed, he followed the river until it flowed between some rolling hills. After a half-hour's work, he drew a circle on the map and said the plane was between the river and a little village shown by a dot. Within two days, the TU-22 was found by our ground forces within the circle that Joe had drawn.

Then-president Jimmy Carter confirmed this event in a speech to Emory College students in September 1995: "American spy satellites failed to locate any sign of the wreckage....It was without my knowledge that the head of the CIA [Admiral Stansfield Turner] turned to a woman [sic] reputed to have psychic powers." Carter also commented that the remote viewer "gave some latitude and longitude figures. We focused our satellite cameras on that point and the [lost] plane was there."[18]

But that wasn't all. As summarized by Larry Dossey, MD: "Two years later, McMoneagle described in detail the unique, secret construction of a 500-foot Soviet *Typhoon*-class submarine being built in a concrete-block building, a quarter-mile from the sea, six months before its launch."[19] The head of the Stargate Project, physicist Dr. Ed May, said of McMoneagle: "Joe McMoneagle is a damn good psychic."[20] For his contributions, McMoneagle was awarded a Legion of Merit award from the United States Armed Forces.[21]

Dr. Dean Radin adds that "McMoneagle has been repeatedly tested in numerous double-blind laboratory experiments and has been shown to have an ability to describe objects and events at a distance and in the future, sometimes in spectacular detail."[22]

Are former president Jimmy Carter, Targ, May, Radin, and others, lying about McMoneagle's abilities? Or could they be real? And if so, how can remote viewing be explained unless consciousness is not confined to the physical brain?

CIA reports publicly released in 2017

In January 2017, the CIA released results of studies conducted on Israeli psychic Uri Geller at the Stanford Research Institute.[23] These documents are now publicly downloadable at www.cia.gov and show that Geller did in fact demonstrate psychic abilities in controlled settings in 1973 (see an exemplary CIA document on the following page). Geller was asked to draw what he "saw" with his mind when the experimenters held a randomly selected picture outside of the room in which he sat. The room was double-walled and electrically isolated. The experimenters comment in the CIA document that they ensured the room was sealed: "In our detailed examination of the shielded room…no sensory leakage has been found."[24] Geller was able to successfully describe and draw the pictures that were outside of the shielded room.

EXPERIMENTS - Uri Geller at SRI, August 4-11, 1973

OBJECTIVE

The objective of this group of experimental sessions is to verify Geller's apparent paranormal perception under carefully controlled conditions and to head toward an understanding of the physical and psychological variables underlying his apparent ability.

EXPERIMENTAL PROGRAM

In each of the eight days of this experimental period we conducted picture drawing experiments. In these experiments Geller was separated from the target material either by an electrically isolated shielded room or by the isolation provided by having the targets drawn on the East Coast. We have continued to work with picture drawing tasks in an effort to achieve repeatability so that we could meaningfully vary the experimental conditions to determine the effect of physical parameters on the phenomena. As a result of Geller's success in this experimental period, we consider that he has demonstrated his paranormal perceptual ability in a convincing and unambiguous manner.

Saturday, August 4. Two drawing experiments were conducted this day. In both of these, Geller was closeted in an opaque, acoustically and electrically shielded room. This room is the double-walled shielded room used for EEG research in the Life Sciences Division of SRI. It is locked by means of an inner and outer door, each of which is secured with a refrigerator-type locking mechanism. Figure 1.

The two drawings used in this experiment were selected by randomly opening a large college dictionary and selecting the first word which could reasonably be drawn. The first word obtained in this manner was

- 1 -

For example, one randomly selected picture was of a bunch of 24 grapes. Geller was asked to describe what the picture looked like—based on what he saw with his mind alone—and he said he saw "purple circles." He then proceeded to draw 24 connected circles (shown on the next page).

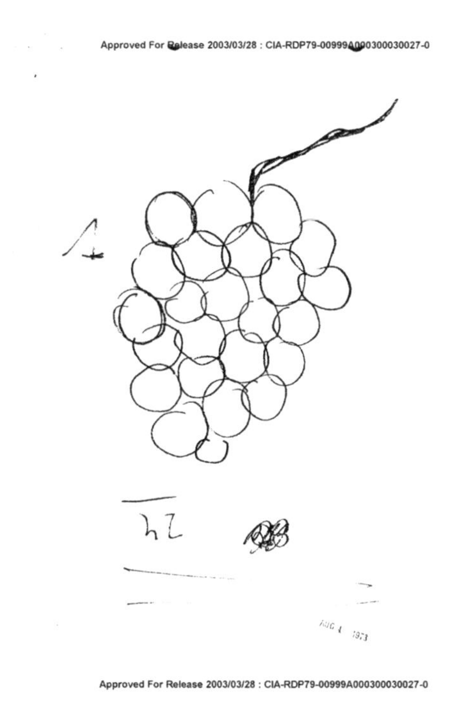

The CIA report's conclusion is clear: "As a result of Geller's success in this experimental period, we consider that he has demonstrated his paranormal perceptual ability in a convincing and unambiguous manner."[25]

Prior to the release of these reports, Geller had been the subject of substantial controversy. Former magician James Randi even wrote a book in 1982 called *The Truth About Uri Geller*, which claimed to debunk Geller's abilities.

However, Targ (who worked directly with Geller) affirmed Geller's abilities, saying: "Many people think that Geller is a total fraud and that he fooled us with his tricks. That is not true. We had more SRI technical and management oversight of our experiments with [Geller] than in any other phase of our research. Hal Puthoff and I found that in carefully controlled experiments [Geller] could psychically perceive and copy pictures that an artist and I would randomly select and draw in an opaque and electrically shielded room.... Geller was an excellent remote viewer."[26]

"Remote viewing is a real phenomenon"[27]

Additional CIA documents discuss remote viewing more generally. They state explicitly that remote viewing is real. See below and the next two pages for extractions from one such document.

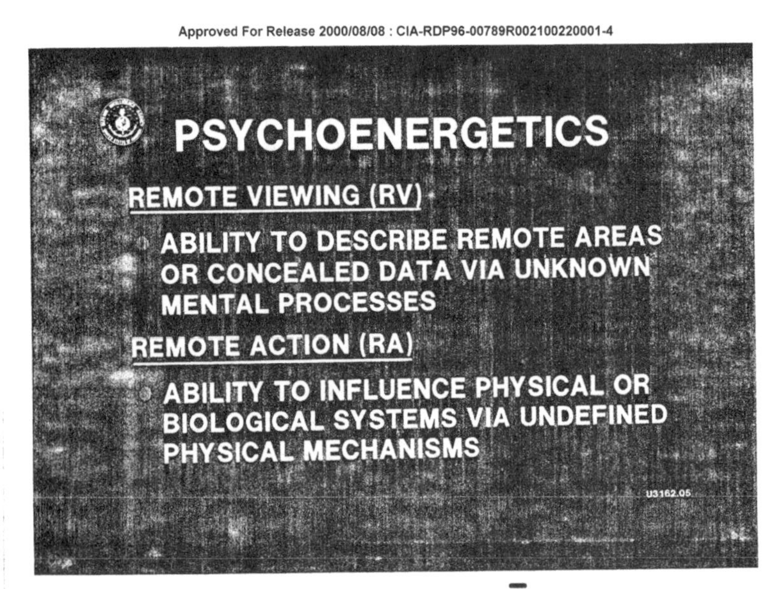

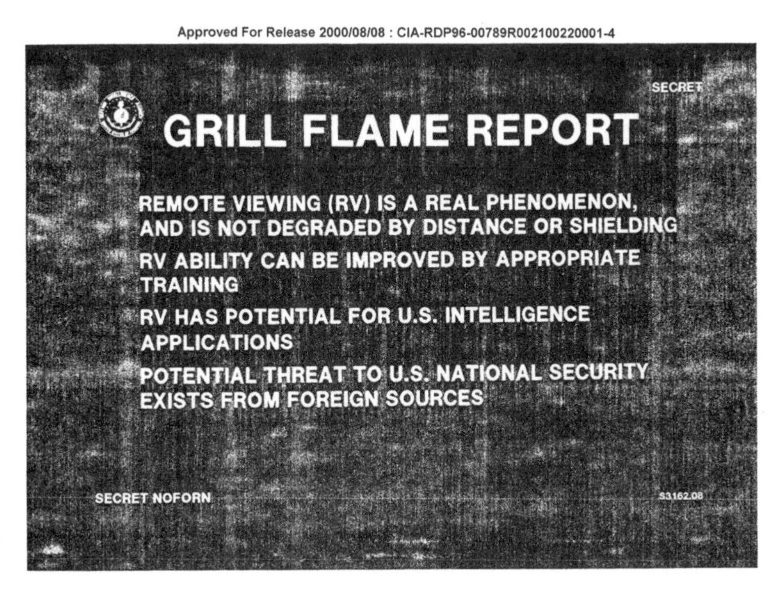
SECRET
GRILL FLAME REPORT
REMOTE VIEWING (RV) IS A REAL PHENOMENON, AND IS NOT DEGRADED BY DISTANCE OR SHIELDING
RV ABILITY CAN BE IMPROVED BY APPROPRIATE TRAINING
RV HAS POTENTIAL FOR U.S. INTELLIGENCE APPLICATIONS
POTENTIAL THREAT TO U.S. NATIONAL SECURITY EXISTS FROM FOREIGN SOURCES
SECRET NOFORN
S3162.08

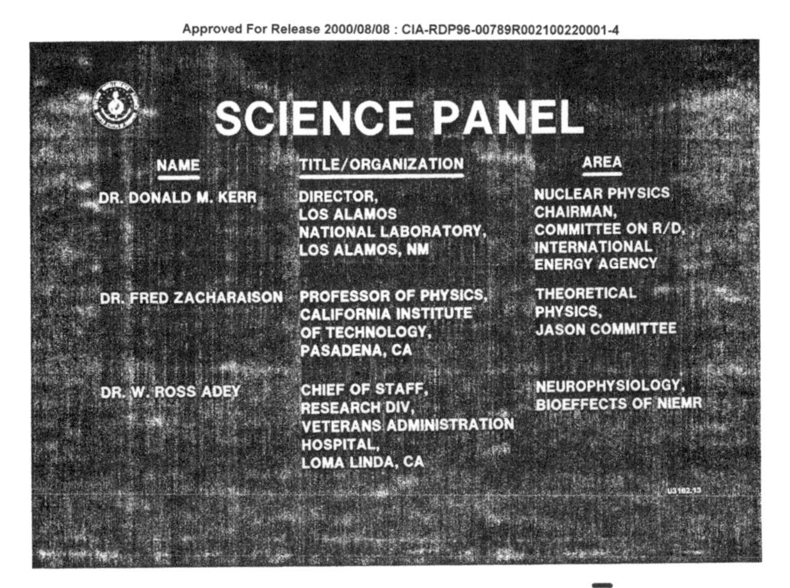
SCIENCE PANEL
NAME
TITLE/ORGANIZATION
AREA
DR. DONALD M. KERR
DIRECTOR, LOS ALAMOS NATIONAL LABORATORY, LOS ALAMOS, NM
NUCLEAR PHYSICS CHAIRMAN, COMMITTEE ON R/D, INTERNATIONAL ENERGY AGENCY
DR. FRED ZACHARAISON
PROFESSOR OF PHYSICS, CALIFORNIA INSTITUTE OF TECHNOLOGY, PASADENA, CA
THEORETICAL PHYSICS, JASON COMMITTEE
DR. W. ROSS ADEY
CHIEF OF STAFF, RESEARCH DIV, VETERANS ADMINISTRATION HOSPITAL, LOMA LINDA, CA
NEUROPHYSIOLOGY, BIOEFFECTS OF NIEMR

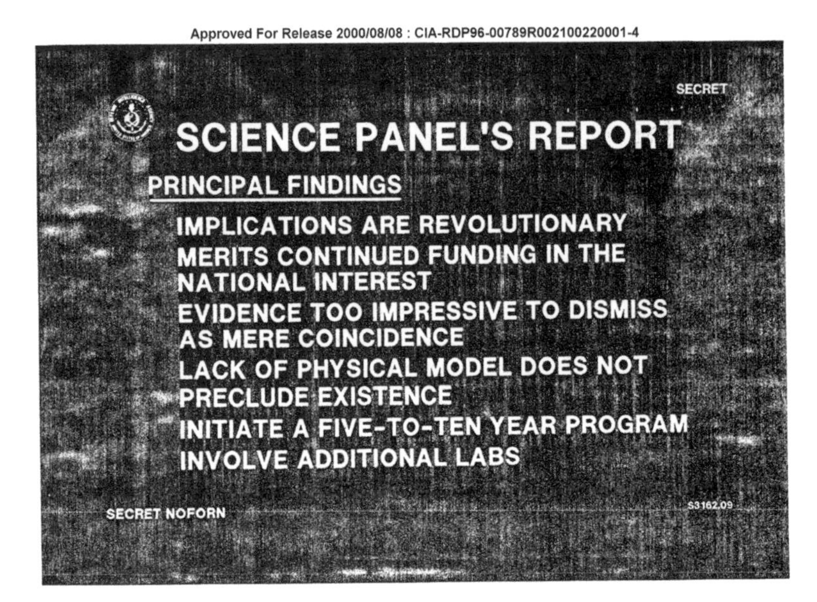

Formal evaluation of the government's program

Upon completion of the Stargate program, two statisticians were asked by Congress and the CIA to evaluate the program's results. One reviewer was Jessica Utts, a statistician from the University of California, Davis, and the 2016 president of the American Statistical Association.[28] In her 1995 review, she states:

> Using the standards applied to any other area of science, it is concluded that psychic functioning has been well established. The statistical results of the studies examined are far beyond what is expected by chance. Arguments that these results could be due to methodological flaws in the experiments are soundly refuted. Effects of similar magnitude to those found in government-sponsored research at SRI and SAIC have been replicated at a number of laboratories across the world. Such consistency cannot be readily explained by claims of flaws or fraud....No one who has examined all of the data across laboratories, taken as a collective whole, has been able to suggest methodological or statistical problems to explain the ever-increasing and consistent results to date.

She also notes, "There is compelling evidence that precognition, in which the target is selected *after* the subject has given the description, is also successful."[29] [emphasis added; more on precognition in chapter 6].

In her assessment of the program's studies from 1973 to 1988, Dr. Utts remarks: "The statistical results were so overwhelming that results that extreme or more so would occur only about once in every 10^{20} such instances if chance alone is the explanation."[30] To be clear, she's saying that the odds that the psychic successes were just luck was 1 in 100,000,000,000,000,000,000. In other words, remote viewing is very likely to be real.

Furthermore, Dr. Utts has shown that the statistical evidence for psychic phenomena is "*much stronger*" than the evidence that aspirin prevents heart attacks[31] [emphasis in original].

A then psychology professor at the University of Oregon, Ray Hyman, a skeptic of paranormal phenomena, was also asked to review the data. After studying Dr. Utts's review, he reported: "[Utts and I]...agree that the...experiments appear to be free of the more obvious and better known flaws that can invalidate the results of parapsychological investigations. We agree that the effect sizes reported in the SAIC experiments are too large and consistent to be dismissed as statistical flukes."[32]

So on top of Dr. Utts's strong statistical results, here we have a renowned skeptic acknowledging that the statistics suggest that remote viewing is real. As Stephan A. Schwartz comments: "This acknowledgment is important because what Hyman is conceding is that the way in which the kinds of laboratory experiments described in the [government] report had been conducted, and the way in which they were analyzed, is no longer a matter for dispute."[33]

Princeton Engineering Anomalies Research Laboratory (PEAR), 1979–2007

The PEAR lab was founded in 1979 by the former dean of engineering at Princeton University, Dr. Robert Jahn, and was co-run by laboratory manager Brenda Dunne. PEAR's general aim was to examine anomalous phenomena of consciousness.

In one set of studies, PEAR examined whether a remote viewer (known as the "percipient") could know where another person (the "agent") would be at some time in the future. The agent's future location was randomly selected *after* the percipient gave his/her psychic impression via remote viewing. In other words, the percipient was asked to predict the future about a place far away.

Here is one example of what a percipient described, 45 minutes *before an agent selected a location* that was 2,200 miles away: "Rather strange yet persistent image of [agent] inside a large bowl—a hemispheric indentation in the ground of some smooth man-made materials like concrete or cement. No color. Possibly covered with a glass dome. Unusual sense of inside/outside simultaneity. That's all. It's a large bowl. If it was full of soup [the agent] would be the size of a large dumpling!"[34]

Not knowing anything about what the percipient described, the agent was randomly assigned to visit the radio telescope at Kitt Peak, Arizona, very much resembling the percipient's description.

In another example, a percipient from Vermont said: "I see (the agent) sitting at a table in an outdoor café or at a brightly lit indoor café table. He is with two or three others, drinking something. [sic] (tea or beer?), and talking. There are leaves and vegetation around, perhaps they're sitting among trees, or there are lots of plants around. The spirit is lively, and the people are having a good time."[35]

The percipient got it right. The agent was in fact at a sidewalk café in Hungary. And the percipient knew this nine days *prior* to the agent's being there. As reported in PEAR's transcript, the agent described the scene: "I went to a sidewalk café with the students at around 1530 hours. We drank beer and wine and sat outside under trees. There were a lot of German tourists around us. I then went to the summer house and drank more beer and wine."[36]

So to recap: The percipient (remote viewer) accurately saw a scene nine days before the agent was at a specific location, 5,000 miles away.

Precisely 653 such sessions were conducted at Princeton. Statistical analysis strongly suggests that remote viewing in these studies was real and that the "hits" were not simply chance occurrences. The statistical odds that the results were due to chance were 33 million to 1.[37]

We've now seen strong evidence for remote viewing from two legitimate sources: the U.S. government and Princeton University. So it's perhaps not totally surprising that even Richard Wiseman, the consulting editor for the *Skeptical Inquirer* publication, concedes: "By the standards of any other area of science[,] remote viewing is proven."[38]

Stephan A. Schwartz

You might be thinking of all the things people could do with remote viewing (if it is in fact real). We'll touch on this in section V when we discuss implications.

For now, consider one more case: Stephan A. Schwartz, former special assistant to the chief of Naval Operations, who has been exploring remote viewing since 1966. He is renowned for using remote viewing to uncover archaeological sites.

As Dr. Larry Dossey summarizes Schwartz's work:

> [Schwartz] is perhaps best known for his role in developing remote viewing…to locate and reconstruct archaeological sites around the world, many of which have eluded discovery for centuries. These include expeditions to Grand Bahama Bank to find the location of the brig *Leander*, to Jamaica with the Institute for Nautical Archaeology to survey St. Anne's Bay and locate the site of Columbus's sunken caravel from his fourth and last voyage; and to Alexandria, Egypt, which resulted in the first modern mapping of the Eastern Harbor of Alexandria and the discovery of numerous shipwrecks. The Egyptian venture also resulted in the discovery of Mark Antony's palace in Alexandria, the Ptolemaic Palace Complex of Cleopatra, and the remains of the Lighthouse of Pharos, one of the seven wonders of the ancient world.[39]

Could this be real?

Under mainstream scientific paradigms today in 2018, remote viewing is impossible and outrageous. Therefore, a skeptic would likely claim that it is not real. But what do we think the odds are that *all* of these cases were fabricated or misinterpreted? Are we to assume that the remote viewers, U.S. government officials, former president Jimmy Carter, laser physicists at Stanford, the former dean of engineering at Princeton, the 2016 head of the American Statistics Association, and Stephan A. Schwartz are *all* lying?

I'm personally skeptical that *all* of these examples could be fabricated.

Chapter Summary

- Remote viewing refers to the ability to "send one's mind" to a distant location at any point in time (past, present, and future), and "see" what's there.
- Laser physicists ran a remote viewing program out of the Stanford Research Institute on behalf of the U.S. government. Certain talented individuals successfully remotely viewed distant objects, and were able to achieve seemingly miraculous feats, such as finding a missing plane in an African jungle (a feat confirmed by former president Jimmy Carter).
- At the request of Congress and the CIA, a renowned statistician examined the data and concluded that psychic functioning appears to exist. A skeptic agreed that the evidence was too strong to dismiss as a fluke.
- Studies at Princeton University, run by the former dean of engineering, also suggest that remote viewing is real.
- Separately, Stephan A. Schwartz has successfully used remote viewing to locate and reconstruct archaeological sites.

Chapter 5

Telepathy
Mind-to-Mind Communication

[The ganzfeld telepathy experiment] has been repeated by dozens of investigators around the world for four decades, including by avowed skeptics who, to their consternation and surprise, successfully replicated the effect.[1]

—Dr. Dean Radin, chief scientist at the Institute of Noetic Sciences

We believe that the replication rates and effect sizes achieved by one particular experimental method, the ganzfeld [telepathy] procedure, are now sufficient to warrant bringing this body of data to the attention of the wider psychological community.[2]

—Dr. Daryl Bem of Cornell University, and Dr. Charles Honorton of the University of Edinburgh

I assume that the reader is familiar with the idea of extrasensory perception, and the meaning of the four items of it, viz., telepathy, clairvoyance, precognition and psychokinesis. These disturbing phenomena seem to deny all our usual scientific ideas. How we should like to discredit them! Unfortunately the statistical evidence, at least for telepathy, is overwhelming.[3]

—Alan Turing, widely regarded as the father of computing, and also responsible for helping to crack German codes in World War II

I'm working presently with several savants who have telepathic abilities.[4]

—Dr. Darold Treffert, psychiatrist, savant specialist, and University of Wisconsin Medical School professor

> *Yes, I think telepathy exists...and I think quantum physics will help us understand its basic properties.*[5]
>
> —Brian Josephson, Nobel Prize-winning physicist

Telepathy refers to mind-to-mind communication. That means communicating without spoken or written words. It means communicating with the mind alone.

Telepathy is often rejected by mainstream science. It is considered fictional. For example, mainstream physicist Michio Kaku states: "True telepathy...is not possible without outside assistance."[6] That might be true under the materialist view of consciousness. But if consciousness is not localized to an individual's body, then telepathy certainly *is* possible.

Nobel Prize-winning physicist Brian Josephson apparently thinks it's real. And pioneering computer scientist Alan Turing thought there was strong evidence. But what did Turing mean in this quotation when he said the "statistical evidence" is "overwhelming"?

Statistical analysis: "effect size"

We touched on statistics in chapter 1, but let's establish some basics here. Statistics can tell us whether an observed phenomenon is likely occurring. Additionally, it can tell us if the phenomenon is likely just occurring by luck. In other words, if we see a statistical effect, then something nonrandom or beyond chance is happening. Something is potentially exerting an effect on the experimental results. In most of the studies referenced in this book, that "something" is consciousness.

A critical fact to understand is that statistical departures from randomness can be small but still statistically convincing. In some of the studies discussed in this chapter and later in the book, we will see effects that are "small but statistically significant." That means something beyond randomness is exerting an effect, but the effect is subtle. Many scientists acknowledge the importance of any statistical effects, no matter how large or small they are.

As stated by Dr. Dean Radin: "It is sometimes imagined that the smaller an effect, the more likely it is due to a mistake rather than a real phenomenon. Who cares about things that are so small that you can hardly see or measure them?...Well, the next time you get a cold virus, which is only about thirty billionths of a meter in diameter, and are in the midst of hacking your lungs out, it may be useful to revisit this question. When dealing with questions of 'Is it real?,' size definitely does *not* matter"[7] [emphasis in original].

In this book, we're trying to understand "Is it real?" So let's keep Dr. Radin's warning in mind.

Furthermore, physicist Dr. Ed May states: "The fact is, that just because it's a weak and small effect, doesn't mean it's unimportant or not real. I mean...some of the most important things in physics and other areas of science have little tiny effect sizes but they are still extremely important."[8]

And because this is such an important topic, here's one more quotation, this time from physicists Bruce Rosenblum and Fred Kuttner. Regarding psychic abilities, they comment: "Any confirmation, *no matter how weak an effect*, would force a radical change in our worldview"[9] [emphasis in original].

With that preface, let's begin our study of the evidence for telepathy by looking at experiments that rely on statistics—where the effects are often small but highly significant.

Ganzfeld experiment

Telepathy has been studied in laboratories for decades. The classic experiment is known as the "ganzfeld" procedure. Remember in chapter 1 when we saw that Carl Sagan said there are three areas of ESP that deserve "serious study"? One of them was "that people under mild sensory deprivation can receive thoughts or images projected at them." He was describing the ganzfeld experiment.

It seems odd that Sagan—a known skeptic—found these studies interesting, and yet Wikipedia's "Ganzfeld experiment" page proclaims: "Consistent, independent replication of ganzfeld experiments has not been achieved."[10] Let's look at the evidence.

In the ganzfeld experiment, one person (the "receiver") sits with half-ping-pong balls over his or her eyes and wears headphones that play relaxing sounds while the experimenter shines a light on the receiver's face. This setup induces a relaxed, dreamlike state. The receiver sits in this relaxed state, in a shielded laboratory room. Meanwhile, another person ("the sender")—who is located in a separate, isolated room—looks at a randomly selected image that the receiver has not seen. The sender is asked to mentally "send" mental impressions of the image to the receiver for a 30-minute period.

Yes, you read that correctly: Someone is trying to "send" a mental image to another person, using the mind alone.

When the period concludes, the receiver is shown four images, only one of which the sender was looking at. The receiver has a one-in-four chance of picking the right image (25 percent odds). Ganzfeld researchers examine whether the receivers do better than 25 percent, in a statistically significant manner. And if they do, the studies would suggest that some sort of mind-to-mind or "telepathic" communication must be involved.

Ganzfeld studies went under the radar for years until 1994, when Dr. Daryl Bem from Cornell University and Dr. Charles Honorton from the University of Edinburgh published results of a meta-analysis of many prior ganzfeld studies. They found that the past studies collectively showed statistical evidence for telepathy with odds against chance of 48 billion to 1.[11] In other words, the odds that the results were occurring by chance were minuscule, meaning the results were likely occurring because of mind-to-mind communication. These results were published in the well-respected *Psychological Bulletin.*[12]

In another meta-analysis of 88 ganzfeld experiments (3,145 trials) conducted between 1974 and 2004, the studies showed "hit rates" of 32 percent rather than the 25 percent expected. Statistically, this 7 percent difference translates into odds against this occurring simply by chance at "29 million trillion to 1." When the study was updated through 2010 to include 4,196 trials, the hit rate was 31.5 percent. Again, this result was above the 25 percent expected. The results translate into odds against chance of "13 billion trillion to 1."[13]

If telepathy was *not* occurring, then we would expect results to approach 25 percent with more and more trials. We would expect the receiver to "guess right" one out of four times. But that's not what the data shows. Instead, the results are consistently 6 to 7 percent above chance. Remember: Just because an effect is small (by one's subjective metrics), that doesn't indicate it isn't meaningful. The results suggest that receivers in a relaxed state with half–ping-pong balls over their eyes are able to receive *some* information that is being sent from another person's mind.

File-drawer accusations

But what if the results were skewed because the researchers withheld the studies that didn't work, only presenting the studies that *did* show an effect? Maybe the researchers were dishonest and hid the results they didn't want people to see. This is a pretty serious accusation to make against experimenters. It implies massive fraud and unscrupulous behavior. However, the accusation is often waged against studies like this and is formally known as the "file-drawer effect" (i.e., the experimenters tucked away the bad results in a file drawer so no one would see them, only publishing the favorable results, and inflating the strength of their data). Since file-drawer accusations have been raised so often, experimenters have been forced to show that the file-drawer effect does not explain the positive results. Experimenters' analyses show that in order for the file-drawer effect to be the explanation for positive results, each of the 30 investigators in the first meta-analysis above would have needed to *not report* 67 additional studies. As Dr. Radin puts it: "To generate this many sessions would mean continually running ganzfeld sessions 24 hours a day, 7 days a week, for 36 years, and for not a single one of those sessions to see the light of day. That's not plausible."[14]

We should also keep in mind that such "questionable research practices" (QRPs) are shown to be most common when more funding for research is available. Research on psychic phenomena (sometimes called "psi") such as telepathy tends to be *under*funded. Psychologist Dr. Imants Barušs and cognitive neuroscientist Dr. Julia Mossbridge elaborate: "Psi research has been relatively poorly funded, with one researcher calculating the total worldwide expenditure on psi since 1882 as equivalent to less than 2 months of spending on conventional psychology research."[15]

Given this trend, file-drawer accusations seem even less likely.

If we reject the file-drawer accusations and accept overwhelmingly statistically significant effects across decades of such studies, then we should conclude that telepathy is real. And remember what that would mean. As biologist Richard Dawkins commented, it would "turn the laws of physics upside down."[16]

This is a huge deal! So huge that we can expect controversy and accusations to continue. But the ganzfeld experiment is only one of many experimental demonstrations of telepathy.

Telepathy in dreams

In a set of studies conducted between 1966 and 1972 by researchers at Maimonides Medical Center in Brooklyn, New York, researchers Montague Ullman and Stanley Krippner examined whether telepathic messages could be transmitted to someone who was asleep.

Participants ("receivers") were placed in a "soundproof dream lab" in which the receiver would sleep with his or her head hooked up to machinery that measured brain waves and eye movements (rapid eye movements are indicative of dream states). Once the receiver's eyes moved rapidly and was assumed to be dreaming, another person ("the sender") would open a sealed package with a randomly selected picture that was selected out of eight pictures. The sender would then try to "send" (with his or her mind) images of the picture to the receiver while the receiver was dreaming. In some cases, the sender and receiver were as far as 45 miles apart.[17]

Once the machinery indicated that the receiver had reached a likely dream state, the receiver was awakened and asked to describe his or her dreams. The receiver's descriptions were recorded and then transcribed. An independent panel of judges—none of whom knew which picture the sender looked at—were asked to rank which of the eight pictures the sender likely saw based on what the receiver described about his or her dream.

The data was clear. According to biochemist Dr. Rupert Sheldrake: "Combining all 450 dream-telepathy trials reported in scientific journals, the overall hit rate was positive and very significant statistically, with odds against this result being due to chance of 75 million to 1."[18] In other words, like the ganzfeld experiments, these "dream" experiments suggest that telepathy is real.

In certain cases, the dreams described were unmistakably close to the image the sender looked at. For example, in one instance the sender was looking at a picture of a boxing match, and the dreamer recalled having a dream "about going to Madison Square Garden in New York and buying tickets for a boxing match."[19]

The sense of being stared at

The evidence for telepathy doesn't end there. People also report sensing that they are being stared at.

In several surveys in the U.S. and Europe, between 70 and 97 percent of people felt that they had experiences during which they knew they were being stared at from behind. [20] Curiously, in occupations in which this matters, professionals are formally trained *not* to stare.[21]

- *Detectives* are told not to stare at the backs of people they are following because the person might turn around and blow their cover.
- *Snipers* report that their targets seem to sense that they're being watched, even if through binoculars. A U.S. sniper in Bosnia who was assigned to shoot terrorists recalls: "Within one second prior to actual termination, a target would somehow seem to make eye contact with me. I am convinced that these people somehow sensed my presence at distances over one mile. They did so with uncanny accuracy, in effect to stare down my own scope."
- *Paparazzi* notice this effect when trying to take pictures of celebrities. For example, one such photographer notes the celebrities would "turn round and look right down the lens.…I am talking about taking pictures at distances of up to half a mile away in situations where it is quite impossible for people to see me, although I can see them. They are so aware it is uncanny."
- *Animal hunters* notice this effect before they are about to shoot their prey. One deer hunter observes: "If you wait a fraction too long, it will just take off. It'll sense you."[22]
- *Martial arts* instructors teach their students "to increase their sensitivity to being looked at from behind."

Intuitively, the sense of being stared at is something many of us can relate to. And it seems like an easily testable phenomenon. Dr. Rupert Sheldrake has done exactly this. He has run studies to examine whether this is a real effect that can be detected under controlled settings.

Dr. Sheldrake's design is simple: one person (the "sitter") has his or her back to another person (the "looker") and is asked whether or not he or she is being stared at. In randomized trials, the looker is either not looking or looking. In some cases, the looker is separated from the sitter by a one-way mirror. In other studies, the two participants are not in the same room, and the looker "stares" by looking at the sitter through a closed-circuit television (CCT). And in some versions of the study, the sitter is wearing a blindfold to eliminate the possibility of visual cues.

The results suggest a strong effect, i.e., that "the sense of being stared at" is real. Dr. Radin summarizes: "I found 60 such experiments involving a total of 33,357 trials from publications cited by Sheldrake and others. The overall success rate in these experiments was 54.5 percent where chance expectation is 50 percent....The overall odds against chance are a staggering 202 octodecillion (that's 2 x 10^{59}) to 1."[23]

Again: we're seeing small, but *highly* statistically significant results. They suggest that the mind is transmitting something that another person subtly picks up. Are our minds "entangled," as Dr. Radin suggests?

Telephone telepathy

Do you ever get a phone call from someone who you were *just* thinking about? You write it off as chance and go about the rest of your day. But you're stunned that before this person called you, you were thinking about him or her.

This is a very common experience. In one study, 92 percent of respondents in surveys all over the world reported experiencing this.[24] Could it be telepathy at work?

Dr. Sheldrake tested this phenomenon more formally. For each participant in the study, his experimenters collected four names and phone numbers of people the participant knew well. The experimenters would then film the participant ("the receiver") as he or she sat in a room alone. At randomly selected times, experimenters would ask one of the four people to call the receiver. Right before the receiver answered the phone, the receiver would say into the camera his or her guess as to who was calling. The phone did not have a caller ID system. The receiver had no known way of knowing who was calling.

The odds of guessing correctly based on pure chance are one in four (25 percent). However, the average hit rate among receivers in the study was 45 percent. This is massively significant from a statistical perspective. It indicates that something well beyond pure chance is happening. Dr. Sheldrake's study has been replicated at universities in Germany and the Netherlands.[25]

Other versions of the same study examined whether the effect is heightened if the caller is emotionally close to the receiver. In these studies, the receiver did much better at guessing on calls from emotionally close callers. The effect was not impacted by how far away in distance the caller was, either. Rather, the effect was just related to how close the person was emotionally.

Similar studies have been conducted with emails and text messages. The experimental results are statistically significant (even when the two communicators are physically far apart) and seem to be impacted by emotional closeness.[26]

Twin telepathy and telesomatic events

Telepathy among "emotionally close" individuals shows up elsewhere: in twins. For example, studies suggest that the extent of "the sense of being stared at" is stronger among twins. They do better in the previously mentioned studies than nontwin siblings. And they do better than people who are unrelated.[27]

However, there are many other cases of telepathic behavior in twins. British psychologist Guy Playfair's book *Twin Telepathy* documents many cases in which telepathy between twins is reported. He stated that his research included "tracking down every reference to the subject of twin telepathy on record since the 18th century that [he] was able to find, interviewing numerous twins and parents of twins, forming a hypothesis, testing it repeatedly, inviting others to replicate it and managing to persuade one of the world's largest twin research units that this is an area that deserves further study."[28]

Did he find that twins are telepathic? He said: "Yes and no; some are and some aren't…somewhere between thirty and forty percent of them are, some of these what we might call occasionally telepathic, invariably at times of crisis, others regularly and a small minority almost permanently so. It is that small minority…that deserve further study. To prove beyond reasonable doubt that telepathy exists, which now seems possible if appropriately funded, would be a discovery worthy of a Nobel prize."[29]

Let's remember: Under the current paradigms of science, *no one* should be telepathic. The fact that *any* twins are allegedly telepathic is indeed significant.

As reported by Playfair, Dr. Lynn Cherkas of the Department of Twin Research and Genetic Epidemiology at King's College in London began noticing many cases of strange, telepathic incidences among twins. She sent a survey to roughly 9,000 twin pairs asking if they experienced "the ability to know what was happening to your partner." Of the 5,513 who replied, more than 50 percent said they were either convinced they had these experiences or thought they "might" have had these experiences. In another survey, 15 percent of identical twins reported experiencing "shared dreams."[30]

One example caught the media's attention. In March 2009, 15-year-old Leanne was taking a bath while her twin sister, Gemma, was downstairs listening to music. Gemma "got the sudden feeling to check on [Leanne]. It was like a voice telling me 'Your sister needs you.'" Gemma checked on Leanne and saved her. It turned out Leanne was having an epileptic seizure and was drowning in the bath. In another instance, Gemma warned Leanne that she would have an epileptic attack. Sure enough, later that day Leanne had an attack.[31]

The mother of a different set of twins wrote to Playfair claiming that 75 to 80 percent of the time, one of her identical twin daughters correctly predicts when the other will have a seizure. "She'll just say, 'Mom, she's going to have a seizure,' or even 'Mom, she's having a seizure' before it even strikes. I have asked her how she knows when a seizure is coming, and she says, 'I just know.'"[32]

Playfair noted that twins are also able to transmit "emotions, physical sensations, and even symptoms such as burns and bruises."[33] Roughly 30 percent of identical twins have such experiences, known as "telesomatic" events.[34] Studies suggest that telesomatic events occur most frequently when an event is negative: "accidents, operations, labour pains, and 'shared negative emotions.'"[35]

In one example, a father accidentally slammed a door on one of his twins' hands. The other twin yelled in pain even though her hand hadn't been smashed. That same twin whose hand *hadn't* been smashed developed a bruise on her hand.[36]

In a similar example, with different twins, one twin "felt 'this sharp pain in my finger' a couple of minutes *before* her sister slammed a door on her finger, [saying] 'I felt it before it happened to her'"[37] [emphasis in original].

Dr. Larry Dossey notes similar examples: "A five-month-old identical twin awakens as the clock strikes ten, and suddenly begins crying. After 15 minutes he stops, as if a switch was turned. At a hospital several miles away, his brother is having a painful injection. His mother notes the time as 10 pm. In a similar report, the mother of another pair of five-month-old identical twins reports that when one of them is having an inoculation he takes it calmly, but the other one 'yells his head off.'"[38]

If one examines these reports through the lens of mainstream materialist science, it would be easy to dismiss them as chance occurrences or as inaccurate accounts. But in the context of an interconnected reality—a nonlocally entangled universe—in which consciousness is not confined to the brain, the stories seem plausible and worthy of investigation.

Telepathy in autistic savants

More extreme telepathy is demonstrated in studies on autistic savants. For example, studies have been conducted by Diane Powell, MD, who earned her medical degree at Johns Hopkins University and is a former Harvard Medical School faculty member. She is currently a practicing neuropsychiatrist. In April 2017 she reported findings of a boy who is able to read his mother's mind when she simply looks at and thinks about randomly generated numbers or words. In these instances, the boy does not see the numbers or words himself. And yet he knows the numbers and words his mother sees.

As Dr. Powell reports: "In June 2016 I tested a 15-year-old autistic boy named Akhil, who types independently and was...[telepathically] accurate. I returned in April 2017 to test him....We used randomized five-digit numbers, words, and nonsense words chosen in advance...and sealed in envelopes before handing them to the mother to open one by one to look at while the boy typed what he 'saw in her mind.' We also tested random words generated by a computer program in real time. His answers contained typing errors, but otherwise were 100 percent accurate."[39]

Whereas the ganzfeld studies, the dream studies, the sense of being stared at studies, and the telephone telepathy experiments were all *statistically significant*, what Dr. Powell shows blows statistics out of the water. She's talking about up to 100 percent accuracy.

This effect is far from being proven, however. Many more replications under tightly controlled conditions will be needed. For example, care will need to be taken to ensure that the child and mother are sufficiently separated so that there is no doubt that the child is not being given cues by the mother. This can be challenging because some autistic children throw a fit when separated from their caretaker.

But Dr. Powell isn't the only researcher to observe this phenomenon. It is perhaps even more significant that a traditionally mainstream psychiatrist like Dr. Darold Treffert (the savant specialist referenced in chapter 2) says he sees telepathic abilities in autistic savants. He states in an August 2017 interview:

> I'm working presently with several savants who have telepathic abilities....When I met these individuals I was very skeptical of that....A number of savants are mute. They simply don't talk. And now with the talking tablet...where they can press out the numbers on the tablet, it gives them a voice....Several patients...have telepathy abilities....[One patient] can read her therapist's mind. It only works between she and her therapist. I'll give the therapist the word card "synesthesia"... or some word that she probably hasn't heard and certainly the patient hasn't heard. And the patient just dutifully types out "synesthesia," a letter at a time....This whole journey has taken me into areas I ordinarily would not have even ventured into.[40]

Coming from a world-leading expert in savants, this is very significant.

Dr. Larry Dossey reports another example: "In one case, George, an autistic savant who could not write his name or a sentence, would know when his parents unexpectedly decided to pick him up at school (he usually rode the bus). He would tell his teacher his parents were coming, and he would be at the door when they arrived."[41]

Biological explanations for telepathy?

Dr. Powell poses a general theory for telepathy from her work with autistic savants. She suggests that these individuals are uniquely skilled at telepathy because their brains are fixed in a state that's innately more "receptive" to picking up forms of consciousness from outside the body. Drawing from the antenna metaphor, one might say that an autistic savant's antenna receiver is better configured to pick up the signal.

Dr. Powell notes in a May 2017 interview that many savants tend to be either autistic or blind, and thus have impairments in certain brain areas relative to normal humans: "Rather than their brains being more connected...you have their brains disconnected in certain areas and those areas are able to work together in a way that's more like a supercomputer." Dr. Powell views this as a "decentralization of intelligence."

She continues: "If you have a model that thinks that memory has to do with the number and complexity of synaptic connections [in the brain] and you're...looking at a brain that has *fewer* of those...and they've got the most phenomenal memory of all, then you have to really question the paradigm"[42] [emphasis added]. Note the similarity between Dr. Powell's statement and the examples discussed in chapter 2, where we saw a relationship between reduced brain activity and heightened experience.

Dr. Powell further notes that an autistic savant's waking brain state is more similar to a nonautistic person's brain state while dreaming. It sounds a lot like participants in the ganzfeld and dream-telepathy experiments. And it also sounds like remote viewers who go into a trance or meditative state when receiving information. Additionally, some psychic mediums, individuals who claim to communicate with the deceased, also go into trance states (see chapter 10).

Maybe savants' brains are simply structured in a way that allows them to somehow pick up "the signal," whereas nonsavants (most of the population) tend to drown out the signal with their logic-dominated brains. This theory would imply that we all have psychic abilities, and we're simply not well trained (or well configured) to perceive them all the time. And therefore it would make sense that savants can be 100 percent telepathic, whereas a nonsavant participant in the ganzfeld experiment does only a few percentage points better than chance.

Further research into these theories is certainly needed. But the collective evidence for telepathy should make us wonder: Maybe consciousness really *isn't* localized to, or produced by, the brain.

Chapter Summary

- "Telepathy" refers to mind-to-mind communication.
- Laboratory studies such as the ganzfeld experiment show strong statistical evidence for the existence of subtle telepathic abilities in everyday people.

- Additionally, studies show that people know when they are being stared at from behind, and they know when someone is staring at them even through a video camera in another room.
- People know beyond chance who is calling them by telephone when the caller is randomly selected. A similar effect is found with email and text messages.
- Many twins are telepathic and know when the other twin is in danger, even when they are apart. Sometimes one twin feels the other twin's pain.
- Anecdotal cases from multiple researchers suggest that some autistic savants are highly telepathic. Much more research is needed in this area before we can draw conclusions.

Chapter 6

Precognition

Knowing the Future Before It Happens

Precognition, in which the answer is known to no one until a future time, appears to work quite well.[1]

—Jessica Utts, University of California, Davis professor and 2016 president of the American Statistics Association, as stated in her 1995 report on psychic functioning (requested by Congress and the CIA)

The accumulated evidence is clear: Precognition exists.[2]

—Dr. Dean Radin, chief scientist at the Institute of Noetic Sciences

We have devoted special attention to this subject...and our collective sense is that true precognition...is a genuine phenomenon.[3]

—Dr. Edward Kelly, professor of psychiatry and neurobehavioral sciences at the University of Virginia and Harvard PhD

The separation that we have between people in space and the separation we have between events in time in the physical world, who says that has to apply to the mental world?[4]

—Dr. Julia Mossbridge, cognitive neuroscientist and leading precognition researcher

Precognition (or presentiment) refers to the ability to know or feel something that will happen in the future. This doesn't make sense based on our ordinary, everyday experience. How could a future event be known if it hasn't happened yet?

But remember what quantum physics teaches us: Time works in mysterious ways. If consciousness is indeed fundamental, perhaps it exists beyond space and time. That would allow for precognition.

Early precognition studies

The early precognition experiments were conducted in the 1930s at Duke University by Joseph B. Rhine. In the basic version of his study, participants were asked to select one of five cards that would later be selected by a random process. Over many trials, participants should "guess right" 20 percent of the time since they have one in five odds of being correct. But over many trials, the hit rate was around 29 percent.[5]

These studies attracted the attention of scientists decades ago. As Pascual Jordan, a colleague of Nobel Prize-winning Wolfgang Pauli, stated in the *Journal of Parapsychology* in 1955: "The existence of psi phenomena, often reported by former authors, has been established with all the exactness of modern science by Dr. Rhine and his collaborators, and nobody can any longer deny the necessity for taking the problem seriously and discussing it thoroughly in relation to its connection with other known facts."[6] This was back in 1955!

Dr. Dean Radin did a comprehensive review of all card tests that were conducted from 1882 to 1939. As he states, for the data "reported in 186 publications by dozens of investigators around the world, the combined results of this four-million trial database translate into tremendous odds against chance—more than a billion trillion to one."[7]

Furthermore, between 1935 and 1987, roughly 309 such studies were conducted by 62 investigators, and 113 articles were published in peer-reviewed journals. More than 50,000 people contributed to roughly two million individual trials.[8] Researchers Dr. Charles Honorton and Diane Ferrari conducted a statistical meta-analysis across all of these studies. As Dr. Dean Radin summarizes: "The combined result showed a small but repeatable effect, with odds against chance of 10^{25} to one. That's ten million billion billion to one."[9]

As we did with the ganzfeld telepathy experiments, we should explore whether the file-drawer effect could be at work here. Did experimenters hide the negative results in a file drawer and not report them so that only the positive results showed? Statistical analysis demonstrates that if this explanation were true, there would need to be 46 unpublished studies for each known, reported experiment.[10] That would imply an unreasonably large number of highly unscrupulous scientists over the course of decades of independent study. This seems unlikely.

If we accept the findings as being accurate, then it appears that somehow people are able to predict future events ever so slightly. They subtly know about the future before it happens. Physics is telling us that time isn't exactly what it seems, so maybe this effect is a symptom.

Skin responses

Dr. Radin examined whether participants' skin conductance would change *before* seeing an emotion-evoking image that they didn't know was coming. He summarizes the experimental design of a study he ran in 1993 (referring to the effect as "presentiment" instead of "precognition"): "While skin conductance is monitored, the participant presses a button. Five seconds later, the computer makes a random decision to display either an emotional [e.g., erotic, violent, or accident scenes] or a calm picture [e.g., landscapes, nature scenes, or calm people]. Presentiment manifests as a rise in skin conductance before emotional pictures, but not before calm pictures."[11]

In order words, Dr. Radin is measuring to see if the body physiologically reacts to a picture, *before* the picture is *randomly* selected by a computer. Yes, before.

In most psychological studies, a stimulus is presented and *then* physiology is measured. Who would ever think to measure the body *beforehand*?

Apparently, the body does in fact respond before the stimulus is presented. The results were clear in Dr. Radin's initial four experiments: "The combined odds against chance for these for four experiments was 125,000 to 1 in favor of a genuine presentiment effect. These studies suggest that when the average person *is about to see* an emotional picture, he or she will respond before that picture appears (under double-blind conditions)"[12] [emphasis in original].

To repeat: Participants' skin was reacting *before* an emotionally provocative image was randomly generated, when the participants did not know the emotionally provocative image was coming.

From 1998 to 2000, Dr. Radin replicated the results in studies at Paul Allen's (cofounder of Microsoft) consumer electronics lab in Silicon Valley, Interval Research Corporation. Biochemistry Nobel laureate Kary Mullis visited the lab and acted as a participant in Dr. Radin's study. Following his participation, Mullis remarked during a 1999 interview on National Public Radio's (NPR) *Science Friday* program:

> I could see about 3 seconds into the future....It's spooky. You sit there and watch this little trace, and about three seconds, on average, before the picture comes on, you have a little response in your skin conductivity which is in the same direction that a large response occurs after you see the picture. Some pictures make you have a rise in conductivity, some make you have a fall. [Radin has] done that over and over again with people. That, with me, is on the edge of physics itself, with time. There's something funny about time that we don't understand because you shouldn't be able to do that...[13]

Is biochemistry Nobel laureate Kary Mullis lying? Is Dr. Radin fabricating the statistics? Or could this be real?

Eye responses

Apparently more than just skin reacts to the future. Dr. Radin measured pupil dilation and levels of spontaneous eye blinking as indicators of arousal before an unknown future event. Again, he found statistically significant results: "This experiment demonstrated that the autonomic nervous system as a whole, reflected in pupil and eye movements, unconsciously responds to future events. That is, it is confirmed that there wasn't anything magically unique about skin conductance measures, as used in the initial presentiment experiments."[14]

So it's not just the skin that's reacting to the future. The eyes react, too.

Brain responses

Psychologist Dick Bierman from the University of Amsterdam examined precognition by measuring brain activity, using a functional magnetic resonance imaging system (fMRI) to examine blood oxygenation levels. He presented participants with a combination of randomly generated erotic images and less arousing images. The results showed a statistically significant effect. According to Dr. Radin: "The brains of both men and women were activated in specific areas *before* erotic pictures appeared, even though no one knew in advance that those pictures were able to be selected. In other words, *the brain is responding to future events*"[15] [emphasis in original].

Dr. Radin also measured electrical signals in the occipital (visual) region of the brain when participants saw randomly flashing lights. A computer generated the lights, so no one could know what the upcoming stimulus would be. Dr. Radin found a statistical effect in female participants: There was more brain activity one second *before* the light flashed.[16]

Heart responses

Dr. Rollin McCraty, director of Research at the HeartMath Institute, and his colleagues examined the heart's role in precognition. Apparently, the heart knows the future, too.

In a 2004 article in the *Journal of Alternative and Complementary Medicine*, Dr. McCraty and his colleagues found changes in participants' heart rates *before* emotional pictures were *randomly* shown. He summarizes: "Of greatest significance here is our major finding: namely...evidence that the heart is directly involved in the processing of information about a future emotional stimulus seconds before the body actually experiences the stimulus....What is truly surprising about this result is the fact that the heart appears to play a direct role in the perception of future events; at the very least it implies that the brain does not act alone in this regard."[17]

The Bem studies

Clearly something is going on—different parts of the body seem to predict the future. The effect is subtle, so we aren't noticeably aware of it.

However, in spite of all this evidence, the effect wasn't getting much attention in the mainstream scientific community. That changed in 2011 when prominent scientist and former Cornell professor Dr. Daryl Bem published his precognition study in the highly respected American Psychological Association's *Journal of Personality and Social Psychology*.

Over a period of eight years, and with more than 1,000 participants in his studies, Dr. Bem found effects of precognition with odds against chance of 73 billion to 1.[18] In other words, as researchers had shown before him, we know the future before it happens.

Dr. Bem remarks about his study: "The research and this article are specifically targeted to my fellow social psychologists.... I designed the experiments to be persuasive, simple and transparent enough to encourage them to try replicating these experiments for themselves." His approach was to "take well-known phenomena in psychology and reverse their time course.... Your physiology jumps when you see [an erotic picture] after watching a series of landscapes or neutral pictures....But the remarkable finding is that your physiology jumps before the provocative picture actually appears on the screen—even before the computer decides which picture to show you. What it shows is that your physiology can anticipate an upcoming event even though your conscious self might not."[19]

The New York Times wrote an article on this topic in 2011 entitled "Journal's Paper on ESP Expected to Prompt Outrage." The author summarizes one of Dr. Bem's procedures: "In one classic memory experiment...participants study 48 words and then divide a subset of 24 of them into categories, like food or animal. The act of categorizing reinforces memory, and on subsequent tests people are more likely to remember the words they practiced than those they did not. In his version, Dr. Bem gave 100 college students a memory test before they did the categorizing—and found they were significantly more likely to remember words that they practiced later."[20]

Bem's paper states: "The results show that practicing a set of words after the recall test does, in fact, reach back in time to facilitate the recall of those words."[21]

Their *future* studying helped their *past* performance on the test. I recognize how strange this sounds.

The New York Times said Dr. Bem's paper "may delight believers in so-called paranormal events, but it is already mortifying scientists."[22] For example, Indiana University cognitive scientist Douglas Hofstadter said, "If any of [Bem's] claims were true, then all of the bases underlying contemporary science would be toppled, and we would have to rethink everything about the nature of the universe."[23] And University of Oregon emeritus professor of psychology Ray Hyman said, "It's craziness, pure craziness. I can't believe a major journal is allowing this work in...I think it's just an embarrassment for the entire field."

Yet the editor of the journal in which the study was published claimed Dr. Bem's paper went through the journal's regular review process, stating: "Four reviewers made comments on the manuscript...and these are very trusted people." He added that all four reviewers decided that the paper met the journal's editorial standards even though "there was no mechanism by which we could understand the results."[24]

A debate continues as to whether the results can be replicated and what a standardized testing methodology should be. However, Bem et al. (2015) conducted a meta-analysis of 90 such experiments from 33 labs located in 14 different countries. The studies contained a mix of successes and failures. But the collective statistics again suggest that precognition is real. The authors note in their meta-analysis: "The fact that exact and modified replications of Bem's experiments produced comparable, statistically significant results thus implies generality across stimuli, protocols, subject samples, and national cultures."[25]

The Mossbridge et al. meta-analysis

Dr. Daryl Bem wasn't the only experimenter to report mind-blowing results on this topic. Cognitive neuroscientist Dr. Julia Mossbridge and her colleagues Dr. Patrizio Tressoldi and Dr. Jessica Utts published a similarly controversial finding in 2012. Mossbridge et al. examined 26 reports between 1978 and 2010, looking at whether the body anticipates ostensibly unpredictable, randomly selected future events. Guess what? Just like we saw in the other experiments, this analysis revealed: "The overall effect is small but statistically significant."[26] For these remarkable findings, Mossbridge et al.'s study won the Charles Honorton Integrative Contributions Award from the Parapsychological Association and was profiled by *ABC News 20/20*, *Wall Street Journal Ideas Market*, *Fox News*, and other media outlets.[27]

Precognitive dreams—anecdotes

The effects described thus far have been subtle. They are only detected by statistics. However, numerous anecdotal cases exist that are much more striking. For example, Dr. Larry Dossey, former chief of staff of the Medical City Dallas Hospital, wrote an entire book on the topic called *The Power of Premonitions* (2009). It was originally inspired by his own premonitions that came to him in dreams, as described in detail below:

> It all began innocently enough, with a dream that occurred during my first year in medical practice. In it, Justin, the four-year-old son of one of my physician colleagues, was lying on his back on a table in a sterile exam room. A white-coated technician tried to place some sort of medical apparatus on

his head. Justin went berserk—yelling, fighting, and trying to remove the gadget in spite of the technician's persistent efforts. At the head of the table stood one of Justin's parents, trying to calm him and lend support. The technician repeatedly tried to accomplish her task but failed as Justin became increasingly upset. Exasperated, she threw up her hands and walked away.

I awoke in the gray morning light feeling shaken, as if this were the most vivid dream I'd ever experienced—profound, numinous, "realer than real." But in view of the dream's content, my reaction made no sense. I did not understand why I felt so deeply moved. I thought about waking my wife and telling her about it but decided against it. What sense would it make to her? We hardly knew Justin, having seen him only three or four times.

I dressed and went to the hospital to make early-morning rounds. As the busy day wore on, I forgot about the dream until the noon hour. Then, while lunching in the staff area with Justin's father, Justin's mother entered the room holding the boy in her arms. He was visibly upset, with wet, unkempt hair and tears streaming down his face. Justin's mom explained to her husband that they had just come from the electroencephalography (EEG) laboratory, where the EEG technician had tried to perform a brain-wave test on the youngster. The tech prided herself in her ability to obtain EEG tracings in children, which can be a demanding task, and her record was virtually flawless—until she met Justin. After relating to her husband how her son had rebelled and foiled the test, Justin's mom left with the disconsolate boy in her arms. Her husband accompanied them out of the dining area and went to his office.

My dream memory returned. I was stunned. I had dreamed the sequence of events in almost exact detail before they happened. I went to see Justin's father in his office and asked him to share with me the events leading up to the aborted EEG. Justin, he related, had developed a fever the day before, which was followed by a brief seizure.

Although he was certain the seizure was due to the fever and not to a serious condition such as epilepsy or a brain tumor, he nonetheless called a neurologist for a consultation.

> The specialist was reassuring; nothing needed to be done immediately…
>
> "Could anyone else have known about these events?" I asked. I wanted to know if someone could have leaked information to me that I might have forgotten, that could have influenced my dream. "Of course not," Justin's dad said; no one knew except the immediate family and the neurologist.
>
> Within a week I dreamed two more times about events that occurred the next day, and that I could not possibly have known about ahead of time.[28]

Dr. Dossey describes one of his personal favorites:

> Amanda, a young mother in Washington State, was awakened one night by a horrible dream. She dreamed that the chandelier in the next room had fallen from the ceiling onto her sleeping infant's crib and crushed the baby. In the dream she saw a clock in the baby's room that read 4:35, and that wind and rain were hammering the windows. Extremely upset, she awakened her husband and told him her dream. He said it was silly and to go back to sleep. But the dream was so frightening that Amanda went into the baby's room and brought it back to bed with her. Soon she was awakened by a loud crash in the baby's room. She rushed in to see that the chandelier had fallen and crushed the crib—and that the clock in the room read 4:35, and that wind and rain were howling outside. Her dream premonition was camera-like in detail, including the specific event, the precise time, and even a change in the weather.[29]

Dr. Dossey also notes that people report seeing winning lottery numbers in their dreams. A recent example: In January 2018, Virginia resident Victor Amole dreamed the numbers "3-10-17-26-32." He claimed, "I've never had a dream like that before." He then purchased four identical tickets with those numbers, which ended up being the winning numbers. He won $400,000.[30]

Precognitive dreams—controlled studies

In addition to the many anecdotal cases reported, there have also been some controlled studies. Caroline Watt, a psychologist from the University of Edinburgh, has been a leader in examining precognitive dreams under controlled conditions. Some of her results are promising. In one study (2014), participants submitted reports detailing their dreams; then they watched a randomly selected video. A group of judges, who did not know which video the participants watched, were asked to judge whether the participants' descriptions of the dreams aligned with the video subsequently shown. Statistical analysis of the judges' evaluations suggests a precognitive effect.[31] In other words, the reported dreams were mildly predictive of the video the participant would watch after the dream. Not all controlled studies on precognitive dreams have shown statistically significant results, however.[32] More data is needed before we can draw definitive conclusions.

Averting disaster

Precognition may also function to save lives. For example, Dr. Diane Powell examined flight-occupancy data on 9/11. The occupancy on each of the four planes that crashed was 51 percent, 31 percent, 20 percent, and 16 percent. The average flight occupancy for a normal plane is 71 percent. Why were there so few people on those planes relative to the average? Did certain people have a sense something bad would happen? One of Dr. Powell's patients did. She was scheduled to be on one of the planes but for unclear reasons decided not to leave Boston. She lived as a result.[33]

In another study, as reported in Michael Talbot's 1991 book *The Holographic Universe*, researcher William Cox studied occupancy rates of trains during 28 serious railway accidents. He found that significantly fewer people took trains on days in which accidents occurred versus that same day in previous weeks.[34]

How can we explain precognition?

Taken together, the accumulated evidence from precognition studies makes one wonder if our mainstream models of consciousness are missing something big. How can we access the future before it happens? Might it truly be that our consciousness exists beyond space and time and therefore can sense "future" events? If so, it's hard to imagine that the materialist model of "The brain creates consciousness" is accurate. That model would expect consciousness to be confined to the present.

One explanation for how this happens is given by statistician Dr. Jessica Utts. She comments on precognitive remote viewing, where people are able to perceive what a distant location will be like in the future. She states: "Precognition, in which the answer is known to no one until a future time, appears to work quite well. Recent experiments suggest that if there is a psychic sense then it works much like our other five senses, by detecting change. Given that physicists are currently grappling with an understanding of time, it may be that a psychic sense exists that scans the future for major change, much as our eyes scan the environment for visual change or our ears allow us to respond to sudden changes in sound."[35]

A fascinating hypothesis, but more research is needed on how precognition might work.

Chapter Summary

❍ Precognition (or presentiment) refers to knowing or sensing that something will happen before it actually happens—in other words, knowing the future.

❍ Studies show that different parts of the body react to future events in laboratory settings, *before* the participants know what's coming:
  — Skin reacts before the future comes.
  — Eyes react before the future comes.
  — The brain reacts before the future comes.
  — The heart reacts before the future comes.

❍ Studies of this nature gained attention in the mainstream in 2011 when prominent Cornell psychologist Dr. Daryl Bem presented findings in a mainstream peer-reviewed journal. His results confirmed that precognition is real.

❍ Mossbridge et al.'s meta-analysis of 26 studies also shows a strong precognitive effect.

❍ Precognition may occur in dreams; sometimes people report dreaming the specifics of an event before the event happens.
  — Precognitive dreams have also been studied in the lab, but the results have been mixed; more data is needed before we can draw conclusions.

❍ Scientists have also observed that some people avert disaster (like 9/11 and train accidents), perhaps by "knowing" the future.

Chapter 7

Animals
Psychic Abilities

> *Through fifteen years of extensive research on the unexplained powers of animals, I have come to the conclusion that many of the stories told by pet owners are well founded. Some animals really do seem to have powers of perception that go beyond the known senses.*[1]
>
> —Rupert Sheldrake, PhD, former Cambridge University biochemist

The psychic phenomena discussed thus far have focused on humans. However, one wonders: If humans can do it, what about animals? If consciousness is fundamental and the brain is simply a lens through which consciousness is experienced, then shouldn't other brained beings (like animals) have similar abilities?

Unfortunately, very little money is flowing into research on animals' psychic abilities. The data that we have is limited, and in this short chapter I will show the highlights. The data that we do have seems worthy of further investigation because it conforms with what we see in humans. We should at least consider the possibility that psychic abilities are at work rather than ordinary senses.

Animals that know when their owners are coming home

Dr. Rupert Sheldrake has been the lead researcher in the field of psychic animals. His 1999 book, *Dogs That Know When Their Owners Are Coming Home: And Other Unexplained Powers of Animals*, is the most comprehensive book written on this topic. One of his most noteworthy findings is that some animals know when their owners are coming home—potentially through nonordinary means.

He states: "Many dogs know when their owners are coming home, cats do it too, and a number of other animals, occasionally rabbits, guinea pigs, quite often parrots and other domestic animals do this. They seem to anticipate the arrival of the person by going and waiting at a door or window, or in the case of parrots they sometimes actually announce verbally who is going to come. They...sometimes know 10 minutes or more in advance. The reason I think it's telepathic is because we've actually done experiments to test this."[2]

Dr. Sheldrake's most famous case is a dog named Jaytee who often knew when his owner, Pam Smart, was coming home. Jaytee typically waited by the window when Pam was on her way home, but rarely spent time there otherwise.[3] A five-minute video example is available on Dr. Sheldrake's website, and it seems to show a clear telepathic effect.[4]

The study's design is as follows:

Cameras are set up at Pam's home to monitor Jaytee's behavior. Pam's parents stay with the dog at home while Pam is driven by the experimenter to a nearby town. Only the experimenters know when Pam will be coming home: Pam doesn't know, her parents don't know, and the cameraman watching the dog doesn't know. While Pam is gone, Jaytee rests by the feet of Pam's mother at the house. Both cameras are time-stamped so one can see Jaytee's activities in relation to Pam's. Within eleven seconds of the experimenter's informing Pam that it is time for them to find a taxi and drive home, Jaytee awakens from his sleep and walks to the window, where he stands waiting until Pam arrives. Pam hadn't even gotten in the taxi when Jaytee reacted. It appears that once Pam had the *intention* of going home, *then* Jaytee walked to the window. She mentally *decided* that it was time to go home, and it was that decision that seemingly triggered Jaytee's movement to the window. It is as if Jaytee read Pam's mind.

The results are difficult to explain by ordinary means. Pam's trip was not part of her normal schedule; she wasn't driving her car—she was far away. She hadn't even gotten in a car when Jaytee reacted, and no one at the house knew when Pam was coming home.

The results have been replicated. Dr. Sheldrake ran roughly 200 experiments with Jaytee and Pam and showed a strong statistical effect, which suggests that animal-human telepathy is in fact real. His results were published in the peer-reviewed *Journal of Scientific Exploration*.[5]

The Jaytee controversy

With such incredible and groundbreaking results, why isn't this talked about more?

If you Google this topic, you might find that the results were debunked by Richard Wiseman, consulting editor of the *Skeptical Inquirer*.

The tale is worth discussing because it highlights the ongoing battle between experimenters and skeptics. Here's what allegedly happened. Wiseman did not believe Sheldrake's results, so Sheldrake allowed Wiseman to run studies on Jaytee and Pam himself (1995). Wiseman provided Sheldrake with the data from his four trials with Jaytee and Pam, and Wiseman's data *matched* the large body of data that Sheldrake had collected!

Then why is there controversy? Apparently, Wiseman did not *report* that he got the same results as Sheldrake, even though he did. Wiseman dismissed Jaytee's abilities in his book *Paranormality*. Additionally, in Sheldrake's words, Wiseman "announced to the world through press releases that they had *refuted* Jaytee's abilities"[6] [emphasis in original].

Sheldrake says Wiseman later admitted that his public refutations were not accurate. As Sheldrake reports: "It was only in 2009 that Wiseman finally conceded that his results showed the same pattern as my own." But the damage had already been done to Sheldrake's reputation.

Sheldrake recalls the story:

> Wiseman reiterated his negative conclusions in a paper in the *British Journal of Psychology*…in 1998. This paper was announced in a press release entitled *Mystic dog fails to give scientists a lead*, together with a quote from Wiseman: "A lot of people think their pet might have psychic abilities, but when we put it to the test what's going on is normal not paranormal." There was an avalanche of skeptical publicity, including newspaper reports [in 1998] with headlines like "Pets have no sixth sense, says scientists" (*The Independent,* August 21) and "Psychic pets are exposed as myth" (*The Daily Telegraph*, August 22). [Wiseman's assistant, Matthew Smith] was quoted as saying, "We tried the best we could to capture this ability and we didn't find any evidence to support it." The wire

> services reported the story worldwide...Wiseman continued to appear on TV shows and in public lectures claiming he had refuted Jaytee's abilities. Unfortunately, his presentations were deliberately misleading. He made no mention of the fact that in his own tests Jaytee waited by the window far more when Pam was on her way home than when she was not, nor did he refer to my own experiments.[7]

So by the time Wiseman admitted that he *did* replicate Sheldrake's results, the public damage had been done.

Dr. Dean Radin looked at the situation closely. After examining the data, he sided with Sheldrake. He remarked:

> Sheldrake...conducted some two hundred test sessions with [Jaytee] that demonstrated to a very high level of certainty that Jaytee did indeed anticipate Smart's return in ways that could not be explained through ordinary means. By contrast, Wiseman conducted a mere four test sessions and yet he claimed that there was no evidence at all to support Jaytee's ability. Ignoring Sheldrake's much larger database was questionable, but even more troubling was the fact that Wiseman obtained exactly the same pattern of results as Sheldrake did, which he failed to mention. In 2002, I asked Sheldrake if I could see the data from his and Wiseman's experiments, and he kindly supplied it. I was able to confirm that the results were indeed strongly in favor of Sheldrake's conclusions, namely that Jaytee's behavior changed dramatically when Pam was returning home from many miles away, and when the signal to return was determined randomly and was unknown to anyone near Jaytee. Sheldrake's designs had evolved to exclude all known loop-holes, and the observed results remained the same. Nothing about this truly exceptional side of the story is mentioned in Wiseman's book.[8]

The implications of the Jaytee studies are profound. If one watches the video on Sheldrake's website, it is difficult to explain by means other than through telepathy or fraud. It is perhaps not surprising that the results have faced challenges in the public eye—they imply the need for a paradigm shift in science.

More broadly, this case is illustrative of the types of challenges that consciousness researchers face. It has been the unfortunate reality for a long time, but hopefully the mounting evidence will begin to change the dynamic.

The examination of animals doesn't end here, however. Other animals exhibit potentially telepathic behavior.

Entangled horses

British horse trainer Harry Blake became convinced, through his many interactions with horses, that they are telepathic. So he conducted a seemingly easy and replicable study. First he separated two sibling horses that had been spending time together. In Blake's design, the horses could not see or hear each other. Blake then observed the behavior of both horses under different scenarios. For example, he would feed the horses at different, irregular times. In 21 out of 24 cases in which one horse was being fed but not the other, the horse not being fed "became excited and demanded food, though it could not see or hear the first one." In other cases, Blake would show lots of attention to one horse, and the other horse, separated from the first, acted as if it was disturbed or jealous.

Blake ran a total of 119 experiments and got positive results in 68 percent of them, whereas in the control group, only 15 percent had positive results. Unfortunately, the studies have not been repeated.[9] The results seem strikingly similar to the telepathic behavior seen among twins, as discussed in chapter 5. If human twins are telepathic, why couldn't horses be telepathic?

Pets finding their owners far away

In some striking cases, pets get lost and find their owners many miles away. It is difficult to imagine that the traditional senses such as smell or hearing can account for these staggering results. Some mechanism of nonlocal consciousness should be considered as a possible explanation.

Dr. Sheldrake has collected a database of such cases. Additionally, researchers at Duke University studied this phenomenon in the 1960s. Dr. Sheldrake reports, "The most remarkable cat story centers around Sugar, a cream-colored Persian belonging to a California family. When they were leaving California for a new home in Oklahoma, Sugar jumped out of the car, stayed for a few days with neighbors, and then disappeared. A year later the cat turned up at the family's new home in Oklahoma, having traveled more than 1,000 miles through unfamiliar territory. Sugar was recognizable not only by her appearance and familiar behavior but also by a bone deformity on her left hip."[10]

Could this have happened by chance? Did Sugar smell or hear her owners from so far away? Or is this another example of telepathy?

In a similar case recalled by Dr. Sheldrake, "Tony, a mixed-breed dog belonging to the Doolen family of Aurora, Illinois, was left behind when the family moved more than 200 miles to East Lansing, around the southern tip of Lake Michigan. Six weeks later Tony appeared in East Lansing and excitedly approached Mr. Doolen on the street. The rest of the family recognized Tony too, and he them. His identity was confirmed by the collar, on which Mr. Doolen had cut a notch when they were in Aurora."[11]

Natural disasters

Like humans (as seen in the previous chapter), animals sometimes know the future before it happens. They seem to know when earthquakes or other disasters are about to strike. For example, as quoted by Dr. Sheldrake: "Before the Agadir Earthquake in Morocco in 1960, stray animals, including dogs, were seen streaming from the port before the shock that killed 15,000 people."[12]

In another case, British biologist Rachel Grant was studying mating behaviors in toads in Italy and noticed that the number of male toads in the breeding group fell from 90 to 0—and this was during mating season. Grant commented: "This is highly unusual behavior for toads; once toads have appeared to breed, they usually remain active in large numbers at the breeding site until spawning has finished."[13] Something seemed fishy. Lo and behold, six days later there was a 6.4-magnitude earthquake. Similar behavior in animals has been observed before tsunamis, avalanches, and air raids.[14]

How do animals do this? Do they somehow "know" the future before it happens? Or is there some other explanation? Regardless of the explanation, science should delve into this topic. Lives could be saved by the development of warning systems, for example (more on this in chapter 13).

A harbinger of death?

Another noteworthy precognition case involves a cat named Oscar who lived in a dementia unit in Rhode Island. He seemed to know when people were about to die. His normal behavior was unfriendly, but several hours before someone would die, he would sit "beside them until they die, often purring and gently nuzzling them." Oscar was more accurate at predicting death than the doctors! Is it possible that he was using psychic abilities rather than ordinary senses?

For instance, as Dr. Sheldrake tells it, "On his thirteenth correct call, a doctor thought one patient was about to die: She was breathing with difficulty and her legs had a bluish tinge. Oscar did not stay in the room with her, so the doctor thought the cat had finally gotten it wrong. But to the doctor's surprise, the patient lived for another ten hours; Oscar returned to join the woman for her last two hours."

In 2007, Oscar was mentioned in the *New England Journal of Medicine* by Dr. David Dosa, who stated: "Since he was adopted by staff members as a kitten, Oscar the Cat has the uncanny ability to predict when residents are about to die. Thus far he has presided over the deaths of more than twenty-five residents....His mere presence by the bedside is viewed by the physicians and nursing home staff as an almost absolute indicator of impending death, allowing staff members to adequately notify families."[15]

Psychokinetic chickens and rabbits

A final example that we will cover delves into psychokinesis—the ability of mind (via intention) to influence physical matter. In the next chapter we will discuss the evidence for psychokinesis in humans. But apparently there is some evidence in animals.

French researcher Dr. René Peoc'h ran studies on baby chickens in the 1980s and 1990s.[16] Baby chickens bond with their mothers at a young age. But if the mother is absent, the chicken will bond with something else nearby (even if it is inanimate).

In Dr. Peoc'h's study, he exposed newly hatched chickens to a small robot so that they would imprint on, and bond with, the robot. The robot moved randomly, as determined by a random- number generator (RNG). RNGs, for example, spit out random sequences of 0's and 1's. So we would expect the robots that move based on RNG outputs to do so in a completely unpredictable fashion.

Dr. Peoc'h had two versions of the study: one in which he observed the robot's movement when there was no baby chicken in a nearby cage (the control), and another in which a baby chicken was in a nearby cage. In the latter case, the chick was able to see the robot. The chick formed an emotional bond with the robot.

When the chick was in a nearby cage, the robot moved *non*randomly. The robot spent a significant amount of its time near the chicken's cage (pattern B below). It was as if the chicken's emotional desire for the robot to be nearby influenced the robot's movement. When there was no chicken in the cage, the robot moved in a random fashion (pattern A below). The studies suggest that chicks can exert influence on the behavior of RNGs using their mental intention alone.

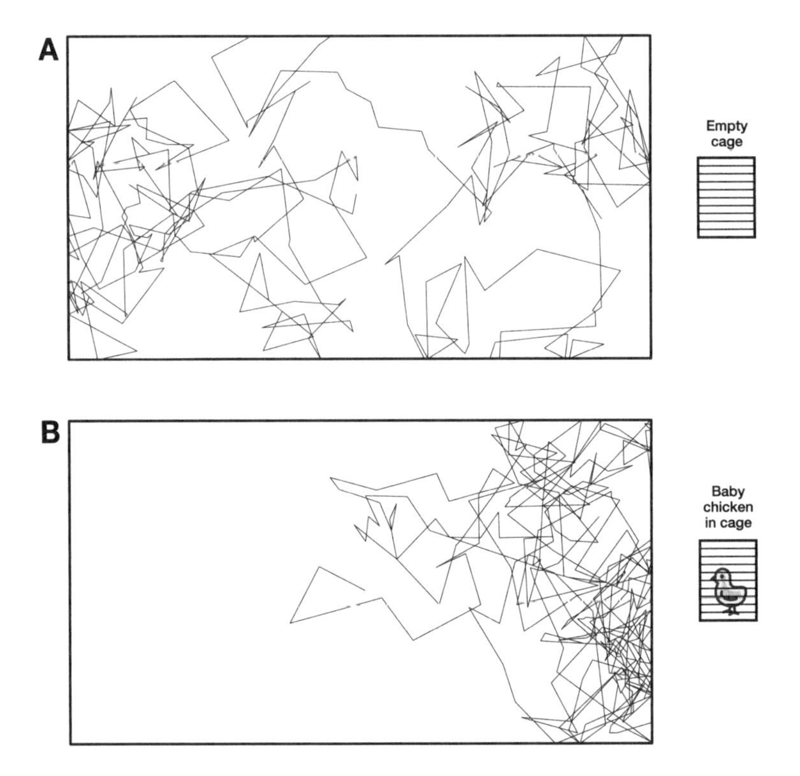

The figure above traces the movement of a small robot in Dr. Peoc'h's studies. In A, the cage is empty. The robot moves in a random fashion because its movement is governed

by a random-number generator (RNG). In B, a baby chicken is in a nearby cage. The robot spends more time near the cage when the baby chicken is present. The results suggest that the chick is exhibiting a mental influence on the robot's random number generator, i.e., a psychokinetic effect. (Figures reproduced from Dr. Rupert Sheldrake's *Dogs That Know When Their Owners Are Coming Home,* page 297.)

Dr. Peoc'h also ran these studies on rabbits. At first, the rabbits were frightened. In those instances, the robot's movement was *away* from the cage. Over time, the rabbits became less afraid and perhaps became attached. The robot then seemed to be pulled toward the rabbits as we saw with the baby chickens.[17]

The results suggest that the nature of the animals' mental intentions can influence the behavior of an otherwise randomly moving robot. Dr. Sheldrake does not know of anyone who has repeated the studies, and certainly independent replications are needed before we can draw conclusions.

However, as we will see in the next chapter, there are *many* studies (including those run by Princeton University's former dean of engineering) which suggest that humans are able to impact RNGs with their minds. If that is indeed true, then why couldn't animals do it?

Chapter Summary

❍ Dr. Rupert Sheldrake ran roughly 200 trials, under controlled conditions, with a telepathic dog who knew when his owner was coming home. The results were replicated by skeptic Richard Wiseman (though it took Wiseman a few years to admit it).

❍ As we saw with human twins, sibling horses seem to exhibit telepathic behavior.

❍ Some animals find their owners when the owners are far away. Many cases are difficult to explain via conventional means.

❍ Animals seem to know when natural disasters are about to occur.

❍ A cat was written up in the *New England Journal of Medicine* for its ability to predict the death of patients in a dementia unit.

❍ In studies conducted by Dr. René Peoc'h, chickens and rabbits apparently directed the movement of random number generator-controlled robots using their minds alone.

CHAPTER 8

Psychokinesis
Mind Impacting Physical Matter

By any reasonable statistical criteria, the likelihood of this concatenation of [psychokinesis] results occurring by chance is infinitesimally small.[1]
—Dr. Robert Jahn, former dean of engineering at Princeton University, and PEAR laboratory manager Brenda Dunne

There is no question that [the Global Consciousness Project] experiment has demonstrated a statistical anomaly.[2]
—Dr. Dean Radin, chief scientist at the Institute of Noetic Sciences

He stood at one corner of the restaurant and he simultaneously bent the spoons of all the people who were there.[3]
—Israeli Prime Minister Benjamin Netanyahu recalling an instance of Uri Geller's spoon-bending abilities

For the last four hundred years, an unstated assumption of science is that human intention cannot affect what we call "physical reality." Our experimental research of the past decade shows that, for today's world and under the right conditions, this assumption is no longer correct.[4]
—Dr. William Tiller, former head of the Material Sciences Department at Stanford University

Most of the evidence from chapters 4 through 7 suggests that consciousness might be nonlocal to the brain. But, aside from Peoc'h's chicken and rabbit studies, we have not looked at whether consciousness can have an effect on the physical world. Is there an "energy" related to intentions and thoughts? Classical physics would say, "No way." Mainstream materialist science says that consciousness is a by-product of brain activity and doesn't interact with physical reality.

But the science of psychokinesis—the impact of the mind on physical matter—continues to challenge that assumption.

If consciousness is the fundamental medium of reality, then we might *expect* the mind to have such powers.

Princeton Engineering Anomalies Research Laboratory (PEAR), 1979–2007

We discussed the PEAR lab in our discussion of remote viewing in chapter 4. The lab was run by Princeton's former dean of engineering, Dr. Robert Jahn, and was managed by researcher Brenda Dunne. In addition to studying remote viewing, PEAR studied psychokinesis.

PEAR found that mental intention *can* impact physical processes, such as the outcome of random number generators (RNGs). As mentioned in the previous chapter, RNGs produce 0's and 1's from random processes. For example, the 0's and 1's might be dictated by noise of electric diodes or radioactive decay.[5] The pattern of 0's and 1's produced should theoretically be completely random and unpredictable.

But let's remember, in Carl Sagan's 1996 book, *The Demon-Haunted World*, he said this area deserved "serious study": "There are three claims in the ESP field which, in my opinion, deserve serious study: (1) that by thought alone humans can (barely) affect random number generators in computers...."[6]

Participants in the PEAR studies are simply asked to focus their mental attention on the RNG. They are asked to *think* about making the RNG produce more 1's than 0's, for example. It's similar to asking someone to influence the frequency of "heads" or "tails" in a coin toss.

Some participants in these studies are often not even onsite. Physicist Dr. Claude Swanson states that most participants "are typical Americans who have lives and homes in other parts of the country. At certain times of day, by agreement, they will turn their mental attention to the [RNG], which is back at the Princeton Lab, and put in a session. Over a twenty-year period, the PEAR Lab has accumulated hundreds of thousands of sessions this way."[7]

A version of the study was conducted with 33 operators over a seven-year period on two different machines. The results: The RNGs behaved nonrandomly when participants focused their mental intentions on the RNGs. Even though the effect was small, it was repeatedly an effect beyond what is statistically predicted by chance.[8]

Interestingly, mainstream physicist Michio Kaku acknowledges this effect in his 2008 book, *Physics of the Impossible*. He states: "PEAR conducted thousands of experiments, involving over 1.7 million trials.... The results seemed to confirm that the effects of psychokinesis exist—but the effects are quite tiny."[9] (Ironically, this concession comes from a chapter in his book on psychokinesis in which he suggests that psychokinesis is problematic because "it does not easily conform to the known laws of physics."[10])

Remember: An effect—even if it is a small effect—is significant. To repeat a quotation from physicists Bruce Rosenblum and Fred Kuttner: "Any confirmation [of psychic phenomena], *no matter how weak an effect*, would force a radical change in our worldview"[11] [emphasis in original].

What's even stranger is that people can impact the *past and future behavior* of RNGs using their minds.[12] Truly remarkable. Dr. Swanson comments: "Of course the whole idea of using intention to affect a radioactive random process...flies in the face of all known physics. The idea of doing it two thousand miles away is even more of an insult to current physics. But to top it all off, the PEAR Lab often asks the operator to make YESTERDAY's results different. And worst of all, for those who want to maintain the present scientific paradigm, they are getting positive results. They are finding that, even under these strange and unlikely circumstances, mental intention does affect the random process."[13]

No wonder Sagan was interested in this.

Brenda Dunne summarizes PEAR's findings:

> We were asking in these experiments whether human intention can actually affect the physical behavior of various kinds of engineering devices. By pure intentionality I mean there are no physical connections, no sensory contact between the human operator and the machines. The question is whether our consciousness can interact with these devices to make a difference in a physical process that can be measured and

> evaluated by standard scientific terms....If you look across all the experiments we've done in the last twenty-seven years, and do a kind of meta-analysis, we're talking about the likelihood of these effects being due to chance as a few parts per billion. What does that mean in layman's terms? It means that [the] effects are very probably real.[14]

Dr. Jahn and Dunne also noticed a performance-related pattern in these studies. When participants relax, they do best. When they are "trying hard," they do less well. Dunne states: "It seems that the less you try to connect with the [RNG] machines, the more successful you are....At first, most people get the effect, however small. Then it almost always reduces significantly. If they try harder and harder, they become frustrated and the effect remains elusive. If they just let it happen, relax, have fun, and gently encourage and tease the machine into cooperating, the effect returns and is often more pronounced."[15]

Group consciousness

If one individual's mental intention can affect physical objects, what happens if a group of people has the same intention at the same time? That is what the Global Consciousness Project sought to (and continues to) examine, starting in 1997.

The project is run by Roger Nelson, PhD, formerly of the Princeton PEAR lab. He set up RNGs around the world that measure patterns of 0's and 1's at any given time. He finds that when major events occur, RNGs behave nonrandomly. In these studies, he's simply measuring the RNGs themselves *without asking people to focus their intentions on the machines*. Most people don't even know that the RNGs are set up. Dr. Nelson is simply testing whether the machines behave differently when people around the world collectively focus their attention in a certain direction—regarding topics completely unrelated to the RNGs.

The Global Consciousness Project provides a concise summary of its setup and mind-blowing findings:

> When human consciousness becomes coherent, the behavior of random systems may change. Random number generators (RNGs) based on quantum tunneling produce completely unpredictable sequences of 0's and 1's. But when a great event synchronizes the feelings of millions of people, our network of

> RNGs becomes subtly structured. We calculate one in a trillion odds that the effect is due to chance. The evidence suggests an emerging noosphere or the unifying field of consciousness described by sages in all cultures.
>
> The Global Consciousness Project is an international, multidisciplinary collaboration of scientists and engineers. We collect data continuously from a global network of physical random number generators located in up to 70 host sites around the world at any given time. The data are transmitted to a central archive which now contains more than 15 years of random data in parallel sequences of synchronized 200-bit trials generated every second.
>
> Our purpose is to examine subtle correlations that may reflect the presence and activity of consciousness in the world. We hypothesize that there will be structure in what should be random data, associated with major global events that engage our minds and hearts.
>
> Subtle but real effects of consciousness are important scientifically, but their real power is more immediate. They encourage us to make essential, healthy changes in the great systems that dominate our world. Large scale group consciousness has effects in the physical world. Knowing this, we can intentionally work toward a brighter, more conscious future.[16]

Some noteworthy examples: RNGs acted nonrandomly during events of global focus, such as midnight at the transition from 1999 to 2000 (Y2K), on 9/11, at the Burning Man festival, and during the days of Pope John Paul II's funeral.[17] The departure of the RNGs from randomness on 9/11 is particularly noteworthy. The departure from randomness began a *few hours before the first tower was hit*. The results make one wonder: Was there a global premonition that 9/11 would occur?[18]

Collectively, the RNG studies imply that the mind can impact the physical world around us. That is not what materialist science teaches us. As Dr. Nelson summarizes the findings: "This is evidence that human consciousness and emotion are part of the physical world.... We interact to produce a mass consciousness even though we are generally unaware that this is possible."[19]

Spoon-bending?

The RNG studies show small, statistical effects that we would never know exist without doing the math. Can psychokinesis occur on larger scales?

A famous example is that of Uri Geller, an Israeli psychic. He demonstrated psychic abilities in government-sponsored studies, and his success was verified in documents recently released by the CIA[20] (as discussed in chapter 4). Geller has been known for his ability to bend spoons using his mind, reportedly having done so at a dinner with former president Jimmy Carter's wife and Henry Kissinger, and also in front of U.S. Army and CIA officials.[21] Israeli Prime Minister Benjamin Netanyahu said in a 2015 television interview that he saw Geller bend spoons with his mind and denies that it was a magic trick: "He did it. And I saw it, and I've seen it time and time again. The fact that you can't explain it doesn't mean it didn't happen."[22] Nobel Prize winner in physics Brian Josephson also endorses Geller's abilities, saying: "I think Uri is a magician, but I don't particularly believe that he is using trickery. I believe there are psychic abilities."[23]

Dr. William Tiller, the former head of the Material Sciences Department at Stanford University, analyzed some of the spoons that Geller claims to have bent. As Dr. Swanson summarizes, Dr. Tiller discovered:

> The metal had an appearance that was different from any kind of material he had ever seen. It was suffused with many tiny regions, each about one thousandth of a millimeter in diameter, which appeared to have been the centers for some kind of local melting of the metal. It had just flowed around these tiny regions in what Tiller called "plastic flow." But there was no evidence that the metal had been heated at all. It was only in these small regions, like tiny bubbles, where the metal seemed to have melted. It was as though energy had been sent into the spot on the spoon which was to bend, but only placed there in these tiny, microscopic hot spots and nowhere else. This phenomenon is unknown in material science. Neither Tiller nor anyone else knows how to replicate this effect. It points out that spoon-bending is a very real phenomenon and that it is based on a mechanism which is unknown to present-day science.[24]

Laser physicist Russell Targ, who confirmed Geller's psychic abilities when conducting research for the U.S. government, says he never saw Geller bend spoons. He "only" saw Geller perform other psychic feats (no big deal).[25]

However, Targ himself now acknowledges that spoon-bending is real. He says: "For two decades I denigrated the whole spoon-bending craze as a kind of silliness. However, a few years ago I saw some metal-bending that changed my mind." Targ became convinced that it's real after attending "spoon-bending parties," during which ordinary people use meditation techniques to bend spoons. He describes one friend who was meditating with a spoon in hand and suddenly it "came alive in her hand and shocked her…she described the experience as suddenly feeling that there was a cricket wiggling against the palm of her hand." The bowl of the spoon had curled "180 degrees toward the handle." Targ also describes successfully bending a spoon himself, emphasizing, "I am not relating these stories to indicate any special psychic prowess on…my part. Rather, I think it is important finally to report there *is* such a thing as paranormal metal-bending, and that it doesn't require Uri Geller to do it. The corollary to this truth is that if we can bend metal at a PK party, then it is quite likely that Geller, who started the craze, can do it also"[26] [emphasis in original].

"Psychoenergetic science"

Large-scale psychokinetic effects have been studied further by Dr. Tiller. In order to explore this alternative realm of science, Dr. Tiller voluntarily left his position at Stanford. He wanted to freely explore the effect of mind on matter, a phenomenon he terms "psychoenergetic science." He gave up the department chair at Stanford, his government and professional committees—all so he would have time to explore this topic. Stanford was not happy with his choice, but Dr. Tiller was a tenured professor and was able to maintain his normal duties while pursuing nonmainstream topics on the side.[27]

Dr. Tiller received outside funding from a philanthropist to support the work. Using the standard experimental protocols of his mainstream Stanford studies, he ran his own studies that successfully showed that the mind can impact matter in significant ways (he makes his technical white papers available for free at his website: https://www.tillerinstitute.com/white_paper.html).

Dr. Tiller's experiments will need additional independent replication before they can be fully accepted. Without further investigation, we are left to question whether these effects are real. However, the preliminary results are striking and deserving of further study.

For example, in his studies, Dr. Tiller "imprints" simple electrical devices with "mental intention." A person focuses his mental intentions on the device in order to imprint it. The imprinted devices then have an impact on the physical world in the direction intended by the mental influencer.

These devices have been used to increase or decrease the pH of water by a full unit—*without* any chemical intervention.[28] That's right—a simple electrical device "stores" someone's mental intention, and that device changes the properties of the physical world around it (in this case, water). This works at a distance too. As Dr. Tiller states: "Intention which we embed from a deep meditative state into a simple electronic device, and that device becomes the host for that intention which we can then ship by FedEx to the laboratory 1,500 miles away or 3,000 miles away or 10,000 miles away probably and they can have the apparatus set up. They just have to plug this into the wall outlet, switch it on and just wait and you will see the properties [of the water] start to change after a month or so."[29]

In other words, people meditate into an electrical device, and that electrical device can then change the pH of water. If this is real, think about what that implies about how powerful our minds are.

Dr. Tiller used a similar method to "condition" a physical space with mentally imprinted devices. Using this method, he was able to increase the chemical activity of a liver enzyme by 25 to 30 percent with a 30-minute exposure to the conditioned space.[30] Furthermore, he used similar methods to speed up development in fruit-fly larvae so that they reached adult stage more quickly. Yes, you read that correctly: he exposed fruit-fly larvae to devices charged with the mental intention to affect chemical processes, which caused the larvae to develop more quickly.[31] So, if this is real, one's mind can alter the development of a living organism.

Given Dr. Tiller's credentials, these studies are worth knowing about. But until the findings are independently replicated on a consistent basis, we need to be careful before jumping to conclusions.

Energy healing

An area that produces similarly noteworthy results, but which also needs more independent replication, is that of "energy healing." Some studies suggest that experienced "energy healers" can focus their mind—their mental intention—to physically heal people. The healer uses his or her mental intention to send energy (sometimes referred to as prana, chi, subtle energy, scalar energy, torsion, or universal life force energy) to whatever is in need of healing. Energy healing methods have been used for thousands of years in Eastern cultures, but are largely discredited in mainstream Western science. The methods sometimes claim to work with the energy centers in the body known as "chakras."

Results of controlled energy healing studies have been mixed. As a result, Russell Targ comments: "Most researchers agree that there is strong evidence for various kinds of distant…healing…*but the results depend strongly on who is doing it*"[32] [emphasis added]. Therefore, he advises: "If you are sick and need a healer, find one who has done it before."[33]

Dr. Joie Jones, professor of radiological sciences at the University of California, Irvine, is an example of a credible scientist who decided to test the outlandish idea that someone's mind could heal another's body. Dr. Jones held a number of impressive positions outside his role in academia. As reported in his obituary in the *Los Angeles Times* (2013), Dr. Jones had "served on President Jimmy Carter's scientific advisory board. He played a similar role in President Barack Obama's first campaign and served as an advisor to the Obama administration on energy and medicine." The obituary acknowledges his work in energy healing: "Jones conducted pioneering controlled experiments that established the healing power of subtle energy."[34]

In his experiments, Dr. Jones placed cells in Petri dishes, exposed them to radiation, and examined the survival rates of cells that received energy healing versus those that did not. In some cases, the healer was thousands of miles away. The results: 50 percent of cells survived when no energy healing was provided by healers, but up to 88 percent survived when the energy healers sent their healing energy. The experiment has been repeated more than 100 times with similar results.[35]

As stated by Dr. Swanson: The effect of the healing energy in this experiment "was not stopped or even reduced by shielding. This proves that the energy responsible for healing is not electromagnetic. This fact, together with the insensitivity to distance, proves that the healing energy is unlike any force known to science."[36]

Another set of studies from the late 2000s, conducted by Pennsylvania State University professor John Neely, MD, also show strong results. Dr.

Neely asked Chinese energy healer Jixing Li (known as a Qigong master) to kill cancer cells. Li was in China and the cancer cells were in Dr. Neely's lab in Pennsylvania. Dr. Neely was asking a man thousands of miles away to kill cancer cells using his mind.

Dr. Swanson summarizes the results: "These experiments show that a Qigong Master can focus his energy very precisely in a small volume, only a few centimeters wide, over great distances of thousands of miles. The energy is capable of killing cancer cells in the targeted region, while sparing those only inches away."[37]

As stated by Dr. Neely: "This is a breakthrough in biological science."[38] Calling this a "breakthrough" is an understatement.

Before we can accept the notion of energy healing as being credible, more research is clearly needed. However, given the apparent strength of evidence for small-scale psychokinetic effects, we do need to consider the possibility that the mind alone can, in some cases, have larger-scale physical effects.

Chapter Summary

❍ Studies over several decades at Princeton University, run by Dr. Robert Jahn (Princeton's former dean of engineering) and PEAR laboratory manger Brenda Dunne show that people can impact the flow of 0's and 1's from random number generators (RNGs)—just by sending their mental intentions to the RNG. Apparently the mind can have an ever-so-slight, but highly statistically significant effect on a physical process.

❍ RNGs set up around the world also behave in nonrandom ways when many people around the world are focused on an event, such as 9/11.

❍ Reputable people claim that spoons have been bent using the mind alone. There are even spoon-bending parties at which people learn to do this.

❍ Some research on large-scale psychokinesis has produced fascinating results, but requires independent replications before we can feel confident that the effects are real.
- Dr. William Tiller, former head of the Material Sciences Department at Stanford, has run studies in which the mind can change the pH

of water, can change chemical activity in a liver enzyme, and can alter the speed of development of fruit-fly larvae.

- Additionally, energy healing experiments suggest that mental intentions can protect cells from radiation and kill cancer cells.

Section IV

Surviving Death? Scientific Evidence

This section provides scientific evidence suggesting that consciousness does not die when the physical body dies. Near-death experiences, communications with the deceased, and children who remember previous lives are examined.

CHAPTER 9

Near-Death Experiences
Lucid Memories with Impaired or No Brain Function

If the prevailing hypothesis, that consciousness is produced by the brain, were correct, there could be no sign of consciousness at the moment when the brain shows no activity. Indeed, this is reported in most cases of clinical death, coma, or brain death. But as the [near-death experience] studies have shown, there are exceptions to this rule. This finding all but forces us to reconsider the relationship between the brain and consciousness. After all, how can people experience an exceptionally lucid consciousness during a period of temporary loss of all measurable brain function?[1]

—Dr. Pim van Lommel, MD, cardiologist

[Near-death experiences] occur in every part of the world, in young people and old. It is critical counter-evidence for the old paradigm theory of consciousness. There is no known physiological mechanism that could account for conscious experience in a nonfunctioning brain. The consciousness that then occurs cannot have been a product of the brain.[2]

—Dr. Ervin Laszlo, systems theorist,
twice nominated for the Nobel Peace Prize

The claim that staunch materialists would like to make is that many of what are called near-death experiences can be explained as residual firing of either the cortex or subcortical structures after it's been damaged or after the heart's stopped.

The problem with that, I call it "wishful thinking," is because if you look at both the human research and animal research, it's well documented that literally all electrical activity ceases within 40 seconds to a minute of the time that bloodflow has stopped going to the brain. There is no evidence of electrical firing of the brain at all. So for people to therefore think that somehow "well, their brain is still firing, it's just that we can't electrically see it"—the reason why they're pushed to that conclusion is because they find it impossible to believe that...consciousness could continue after the brain had stopped functioning. That's because they're stuck believing this simple materialist view.[3]

—Dr. Gary Schwartz, professor at the University of Arizona, former professor at Yale University, and Harvard PhD

No one physiological or psychological model by itself explains all the common features of near-death experiences....The paradoxical occurrence of heightened, lucid awareness and logical thought processes during a period of impaired cerebral perfusion [blood flow to the brain] raises particularly perplexing questions for our current understanding of consciousness and its relation to brain function....A clear sensorium and complex perceptual processes during a period of apparent clinical death challenge the concept that consciousness is localized exclusively in the brain.[4]

—Dr. Bruce Greyson, professor emeritus of psychiatry and neurobehavioral sciences at the University of Virginia

The idea that a consciousness can exist and make memories independent of the brain...is a startling finding.[5]

—Dr. Allan Hamilton, University of Arizona neurosurgery professor and Harvard Medical School graduate, commenting on a patient's detailed memories that occurred during the time of cardiac arrest when there was no measured brain function

You've probably heard the term "near-death experience" (NDE). You might be vaguely familiar with the idea that when some people nearly die—or are clinically dead before being brought back to life—they report having had clear and blissful experiences reminiscent of an "afterlife." They claim they "saw the light." They felt overwhelmed by unconditional love.

And in many cases they came back to life forever changed, caring more about helping the world and less about materialistic goals.

I had heard about experiences like these in passing. I remembered hearing that they were just hallucinations caused by a dying brain. They weren't telling us anything new about the nature of reality.

That was before I looked at the research, which paints a very different picture: It suggests that unexplained conscious experience exists even when the brain is severely damaged or even "off." How can these experiences occur in patients under general anesthesia or in cardiac arrest with no measured blood flow to the brain?

I'll first walk you through the history of NDEs and their common features. Then we'll explore the various unproven theories that claim that the brain explains NDEs. All of the theories suffer from flaws upon closer examination. The most logical explanation of NDEs doesn't include the brain at all. Instead, the evidence suggests that consciousness exists independently of the brain.

So, for a moment, let's suspend disbelief and look at the science.

The basics of NDEs and their history

What exactly is an NDE? The International Association of Near-Death Studies (yes, such an association exists) defines an NDE as: "A profound psychological event that may occur to a person close to death or, if not near death, in a situation of physical or emotional crisis. Because it includes transcendental and mystical elements, an NDE is a powerful event of consciousness; it is not mental illness."[6] We'll get to those "powerful events" shortly.

NDEs are typically prompted by events causing severe brain impairment, such as cardiac arrest, comas from brain damage, near-drowning, and other traumas.[7]

Roughly 13.5 million people in the United States alone have reported having an NDE.[8] Even if we assume that some of those cases were fabricated, that's still a lot. Forums are developing that track NDEs, such as radiation oncologist Dr. Jeffrey Long's nonprofit Near-Death Experience Research Foundation. It has a collection of more than 4,000 reported NDEs in 23 languages.[9]

We should keep in mind that the field of NDE study is relatively new. That's because resuscitation technology (i.e., bringing people back to life) is also relatively new. As described in *The Handbook of Near-Death Experiences: Thirty Years of Investigation* (2009): "By the early 1970s, resuscitation technology had advanced to the point where people were being brought back from the brink of death in numbers unprecedented in human history. The conditions they survived were as extreme as several minutes of cardiac arrest.... For the first time in history, enough people were reporting this type of experience that professional health care providers began to discern *patterns* in the experience itself and in the aftermath of it"[10] [emphasis added]. So it's only in the last few decades that we've seen a rise in NDEs, to the point that they can be closely studied.

The first scientist to study these patterns and popularize NDEs was Raymond Moody Jr, MD, PhD. He put NDEs on the map in 1975 when he released his international bestseller *Life After Life*, in which he summarized firsthand accounts from roughly 150 near-death experiencers (NDErs).

However, even before NDEs became a mainstream topic, the phenomenon was described elsewhere. At the time Dr. Moody published his book in 1975, more than 30 articles in Western periodicals—written by more than 25 different authors—discussed NDEs.[11] Furthermore, in roughly 95 percent of the world's cultures throughout history, some form of near-death (or out-of-body) experience has been reported. Plato's *Dialogues*, *The Egyptian Book of the Dead*, and *The Tibetan Book of the Dead* discuss phenomena that are strikingly similar to what we now call NDEs.[12]

Physicist Dr. Claude Swanson remarks on similarities found across NDEs: "The consistency in the accounts of near-death experiences is one aspect which gives them credibility. These accounts come from people of all ages, cultures, and religions."[13]

Dr. Moody agrees: "What has amazed me since the beginning of my interest are the great similarities in the reports, despite the fact that they come from people of highly varied religious, social, and educational backgrounds."[14]

NDEs transform people's lives

Additionally, people behave similarly after their NDEs. An NDE often "permanently and dramatically alters the person's attitudes, beliefs, and values, often leading to beneficial personal transformations." NDErs tend to become less materialistic, less competitive, have a greater concern for others, appreciate life more, fear death less, and feel a greater sense of purpose.[15]

A noteworthy example is the NDE of Anita Moorjani. She was in a coma, on her deathbed, after a four-year battle with cancer. Her doctor told Moorjani's husband, "There's nothing we can do for your wife, Mr. Moorjani. Her organs have already shut down. Her tumors have grown to the size of lemons throughout her lymphatic system, from the base of her skull to below her abdomen. Her brain is filled with fluid, as are her lungs. And as you can see, her skin has developed lesions that are weeping with toxins. She won't even make it through the night."[16]

But Moorjani had a transcendental experience while in a coma, where she felt immersed in unconditional love; communicated with her deceased father, friends, and others; and realized that she had been unnecessarily harsh and judgmental in her life, spending too much time worrying about the opinions of others. She realized that by shifting her mind-set, she could cure herself of cancer.

Moorjani awoke from her coma, and within two days her organs began functioning. As she recalls it:

> When the oncologist performed a routine checkup, he couldn't hide his surprise: "Your tumors have visibly shrunk—considerably—in just these three days!" About six days after coming out of the ICU, I began to feel a little bit stronger and was starting to walk up and down the hospital corridor for short periods of time before needing to rest. Every day the doctors reported on my latest test results. "I don't understand. I have scans that show this patient's lymphatic system was ridden with cancer just two weeks ago, but now I can't find a lymph node on her body large enough to even suggest cancer," I heard him say. To the amazement of the medical team, the arrangements they'd made with the reconstructive surgeon to close the lesions on my neck were unnecessary because the wounds had healed by themselves. On March 9, 2006, five weeks after entering the hospital, I was released to go home. And I couldn't wait to live my life with joy and abandon![17]

Moorjani is now a best-selling author and speaker, teaching others about her transformative experience.

In other instances, NDErs have trouble adjusting to being back in their physical body. As one NDEr reports: "After my NDE I felt like a child learning how to walk. The world around me overwhelmed me."[18] NDErs also tend to become more psychic and intuitive, and some report seeing colorful "auras" around people following their NDE.[19]

Some NDErs find that their electrical devices, such as wristwatches, stop working. For example, NDE researcher Dr. Jan Holden joked at a 2016 NDE conference that you always know who's an NDEr and who is a researcher at the conference because the NDErs aren't wearing watches.[20] NDE researchers Kenneth Ring and Evelyn Valarino describe the same finding: "A surprisingly large proportion of these persons discover, for instance, that digital wrist watches will no longer work properly for them, or they 'short out' electrical systems in their cars, or computers and appliances malfunction for no apparent reason, and so on."[21]

Typical elements of an NDE

What exactly happens in an NDE that's so commonly experienced and profoundly transformative?

Below is a list of experiences consistently reported, even though any one NDE might not have all of these features.[22]

1. *Ineffability:* NDEs are difficult to put into words. One NDEr states: "I regret that words can't do my experience justice. I must admit that human language is woefully inadequate for conveying the full extent, the depth, and the other dimension I've seen. In fact, no pen can describe what I went through."[23] It is therefore difficult for someone who has not had an NDE to fully relate to an NDEr's account.
2. *Positive emotions:* NDErs report experiencing unconditional love, peace, joy, bliss, a disappearance of pain, and a mystical feeling of oneness.[24] For example, an NDEr reports: "I was immersed in a bluish white light that had a shimmering appearance as if I were swimming underwater in a crystal clear stream.... The visual was accompanied by a feeling of absolute love and peace.... I had fallen into a pure positive flow of energy. I could see the flow of

energy. I could see it flow through the fabric of everything....I remember thinking, 'This is the greatest thing that can ever happen to anyone.'"[25]

Similarly, Anita Moorjani recalls that during her NDE she "felt completely enveloped in a sea of unconditional love and acceptance. I was able to look at myself with fresh eyes, and saw that I was a beautiful being of the universe."[26] While many NDEs are positive experiences, a minority are negative and fear-inducing.[27]

3. *Heightened senses:* In one NDE survey, 74.4 percent of respondents stated they experienced "more consciousness and alertness than normal"[28] relative to their everyday levels. Some people indicate having 360-degree vision.[29]

4. *Acknowledgement of being dead:* In one example, an NDEr reports people trying to save his lifeless body after being struck by lightning, recalling: "I spoke to the people around my body but they could not see or hear me; I could see and hear everything they did and said. It suddenly occurred to me that I was thinking normal thoughts, in the same mental vernacular I had always possessed. At that moment I suddenly had one simple, ineloquent and rude thought, 'Holy shit, I'm dead.'... Interestingly there was no strong emotion accompanying my apparent death. I was shocked....I felt no reaction to what should have been the most emotional of life's events."[30]

5. *Out-of-body experience (OBE):* An OBE is the experience of feeling separated from the body, sometimes looking down at the body from above, and sometimes reporting seeing and hearing events that actually happened during the time of the NDE.

6. *Encountering "heavenly" realms:* An NDEr reports: "The landscape was beautiful, blue skies, rolling hills, flowers. All was full of light, as if lit from within itself and emitting light, not reflecting it."[31] Another recalls: "All around me I could see and feel a beautiful peace and tranquility with love and peace....As far as the eye could see to my left was a beautiful landscape of tulips of every color imaginable."[32]

7. *Experiencing a dark space or tunnel:* One NDEr states: "My next awareness was of being submerged and cradled in a warm, wavy, wafting motion at the opening of a tunnel. The tunnel had

billowy soft sides and was well lit, with the tunnel dimensions decreasing and brightness increasing as it got closer to a single bright light."[33] In one survey, 33.8 percent of NDEr respondents said they passed into a tunnel during their NDE.

8. *Encountering a mystical, brilliant light or "being" of light.* Cardiologist Dr. Pim van Lommel summarizes: "The light is described as an extremely bright, nonblinding light that permeates everything. People are ineluctably drawn to this light and are usually completely enveloped by it. Sometimes this light is experienced as a being."[34]

9. *Encountering either mystical or divine beings or deceased relatives or friends:* For example, NDErs see deceased individuals they recognize. In some cases, as stated by Dr. van Lommel, "These [deceased] people look healthy again, even though the prevailing memory of them is as very sick and weak in the period before they died. If they died at a very young age, they may look like young adults now."[35]

10. *A sense that time or space is different:* An NDEr recalls: "Seeing no point in staying with my body, my thoughts then moved to walking away....I had lost all form entirely and instead was just a ball of energy and thought....Instead of bothering with the stairs, I passed through the wall....I was aware of every moment of this experience, conscious of every millisecond, even though I could feel that time did not exist."[36]

11. *A panoramic "life review":* NDErs report experiencing their lives in a flash. They experience their lives not only from their perspectives, but sometimes from the perspectives of those they affected. As they experience their life reviews, they judge themselves and learn lessons. As reported in one representative life review: "Every second from birth until death you will see and feel, and [you will] experience your emotions and others that you hurt, and feel their pain and emotions. What this is for is so you can see what kind of person you were and how you treated others from another vantage point, and you will be harder on yourself than anyone to judge you."[37] NDErs "realize that every single thought, word, or action has a lasting effect on themselves and others."[38] Sometimes the NDEr is shown "previous lives" during the life review.[39]

12. *Encountering or learning special knowledge; understanding universal order and life purpose:* An NDEr reports: "All the secrets of the universe were revealed to me."[40]

13. *A preview or flash-forward:* "People feel like they can see part of the life that is yet to come,"[41] some of which is verified in the future.

14. *Perceiving a boundary or barrier:* The NDEr is aware that if he or she crosses a certain boundary or barrier, he or she will not be able to return to the body.

15. *A return to the body, either voluntary or involuntary:* One example is an NDEr who was struck by lightning and reports that after feeling a state of bliss, "Suddenly, I was back in my body. It was so painful. My mouth burned and my left foot felt like someone had stuck a red-hot poker through my ankle."[42]

Remember: People who have NDEs are seriously injured or ill. Their brain is either severely compromised or not functioning at all. And yet NDErs consistently describe having this hyper-real pattern of experiences.

As stated by Dr. Long: "Considering NDEs from both a medical perspective and logically, it should not be possible for unconscious people to report highly lucid experiences that are clear and logically structured. Nonetheless most NDErs report supernormal consciousness at the time of their NDEs."[43]

And let's not forget that some NDEs occur when the person is under general anesthesia. In fact, the University of Virginia's database shows that 23 percent of NDEs it documented were of patients under anesthesia, "and these involved the same features that characterize other NDEs, such as having an [out-of-body experience] and watching medical personnel working on their body, an unusually bright or vivid light, meeting deceased persons, and—significantly—thoughts, memories, and sensations that were as clear or clearer than usual."[44] Dr. Long reminds us that according to mainstream neuroscientific assumptions, a brain under anesthesia "should not be able to produce lucid memories."[45]

However, materialist science resists the notion that NDEs give a window into some broader realm of consciousness. Let's look at some of the counterarguments.

Unproven theory: NDEs are caused by expectations

One might wonder: Maybe the similarities in NDE reports are caused by people's expectations. Maybe they've heard about NDEs, which makes them more inclined to have a similar experience. They expect it because they've heard about NDEs on TV or through other cultural means. Dr. Bruce Greyson of the University of Virginia refutes that idea. He notes that many NDE reports come from individuals "who had no prior knowledge about NDEs."[46]

Also, young children have NDEs that fit the standard pattern. Often they are too young to know about NDEs. Some children who have a typical NDE are under the age of five![47] As cardiologist Dr. van Lommel asks rhetorically: "Children who have a near-death experience remember the same typical elements as adults; but how is this possible when children have never heard of near-death experience or, in some cases, have not even learned how to read yet?"[48] Or as Cherie Sutherland, PhD, puts it, following her review of 30 years of NDEs in children: "It has often been supposed that the NDEs of very young children will have a content limited to their vocabulary. However, it is now clear that the age of children at the time of their NDE does not in any way determine its complexity. Even pre-linguistic children have later reported complex experiences.... Age does not seem in any way to affect the content of the NDE."[49]

Physiological explanations

We will now briefly review a number of theories claiming that physiology alone can account for NDEs. However, let's keep in mind that at this point all such ideas remain theoretical. Mainstream materialist science does not yet have a definitive explanation for how NDEs can occur—it cannot adequately explain all of the features of NDEs by physiology alone.

Psychologist Dr. Imants Barušs and cognitive neuroscientist Dr. Julia Mossbridge summarize a core problem with physiological theories: "It looks as though the less the brain is able to function properly, the more vivid the experiences that are occurring, assuming that the experiences are occurring at the same time as the brain is shutting down. On the basis of that assumption, the physiological data showing impairment in brain function during NDEs actually undermine physiological explanations for NDEs."[50]

While physiological explanations suffer from this fundamental flaw, they are still worth examining because they are often referenced.

Unproven theory: NDEs are caused by a lack of oxygen

One example of a physiological explanation is that a lack of oxygen to the brain causes NDEs. Dr. Greyson (among others) disagrees. He first notes that, in general, a person who has a hallucination caused by low oxygen would typically experience "agitation and belligerence." Many NDErs, on the other hand, typically have peaceful, loving experiences. So that doesn't align. On top of that, he cites studies showing that "[NDErs] have oxygen levels the same as, or higher than, those who do not have NDEs."[51] Dr. van Lommel further explains that in one of his studies, all patients were deficient in oxygen and yet not all of them had NDEs. If lack of oxygen was the cause, then he would have expected *all* of the subjects to have NDEs. But they didn't.[52] Lack of oxygen doesn't seem to explain NDEs well.

Unproven theory: NDEs are caused by increased carbon dioxide levels

Others have similarly theorized that NDEs resemble what happens to people when they inhale carbon dioxide. However, science lacks empirical data showing that increased carbon dioxide levels *cause* NDEs. In one study, researchers found a correlation between the intensity of the NDE and carbon dioxide levels.[53] Yet, people who inhale carbon dioxide do not report NDE features such as a life review with specific, accurate life details.[54]

Unproven theory: NDEs are caused by endorphins

Others theorize that NDEs are caused by endorphins that are either produced by the body or are given to the NDEr. However, endorphins are known to produce long-lasting pain relief. On the other hand, NDEs are very brief: "The onset and cessation of the NDE and its associated features is usually quite abrupt, with the pain relief lasting only as long as the NDE itself, which may be only seconds." If endorphins were the cause, we'd expect a longer period of pain relief. Also, endorphins fail to account for common NDE features such as out-of-body experiences, the life review, and encountering deceased individuals.[55]

Unproven theory: NDEs are caused by ketamine

The anesthetic chemical known as ketamine has similarly been theorized to cause NDEs. As neuroscientist Sam Harris, PhD, puts it: "Such compounds are universally understood to do the job"[56] of producing the kind of clarity described in NDEs. University of Virginia professors Dr. Emily Williams Kelly, Dr. Bruce Greyson, and Dr. Ed Kelly take a different stance, citing evidence to the contrary. They note that while ketamine can produce some features of NDEs, ketamine users typically report that the experiences don't seem real. And those experiences are sometimes described as being "frightening" or involving "bizarre imagery."[57] Most NDEs, on the other hand, are described as being enjoyable and seem real. Also, ketamine usually acts on a functional brain, whereas some NDErs have reduced (or no) brain function.[58]

Unproven theory: NDEs are caused by DMT

The hallucinogen N,N-Dimethyltryptamine (DMT) has been referenced as a possible cause of NDEs (e.g., by Dr. Harris).[59] DMT is naturally produced by the body, and it can also be taken as a psychedelic drug. It is reported to cause lucid, mystical experiences and produces a "rich ultra-reality"[60] sensation. Neurosurgeon and former associate professor at Harvard Medical School Dr. Eben Alexander disagrees. Dr. Alexander had a life-transforming NDE during a prolonged coma from bacterial meningitis. During his NDE, the parts of Dr. Alexander's brain responsible for "rich ultra-reality" sensations were not functional. As he explains, "My cortex was off, and the DMT would have had no place in the brain to act."[61] Dr. Alexander also contends: "So in terms of 'explaining' what happened to me, the DMT-dump hypothesis came up…radically short"[62] in trying to account for such a transformative NDE.

And if we adopt the metaphor that psychedelic substances open the brain's "reducing valve," then psychedelics simply enable the brain to experience a broader reality—a reality that is always present but normally obfuscated from our everyday field of perception. They don't create the alternative reality; they are a gateway into it. So even if Dr. Harris is correct that DMT plays a role in some reported experiences, DMT wouldn't necessarily be the true *cause*; it would just be an enabler.

Unproven theory: REM intrusion causes NDEs

Some, such as Kevin Nelson, MD, adopt the "REM intrusion" hypothesis to explain NDEs. REM stands for "rapid eye movement" and occurs during deep sleep. Complex dreaming occurs during REM. "REM intrusion" occurs "in the transitions between REM and waking.…The blending of REM and waking consciousness takes the form of complex visual and auditory hallucinations, [and] dream narratives....This borderland is unstable, lasting seconds or minutes before reverting to a more stable conscious state."[63] Like others, this hypothesis fails to account for the totality of NDE experiences. Dr. Alexander summarizes: "Special features, such as...the occurrence of typical NDEs under general anesthesia and other drugs that inhibit REM, rule against any explanation of NDEs through REM intrusion."[64]

Unproven theory: NDEs are delusions

Still, some scientists insist that NDEs must be made up—they must somehow be hallucinations that the brain created, even if we don't understand how or why. That becomes difficult to argue when studying the data on life reviews that reveal true facts rather than delusions.

Let's first look at a full account of how a life review is typically experienced. If real, this account (and others) are truly remarkable and might make us rethink our lives. As stated by an NDEr:

> All of my life up till the present seemed to be placed before me in a kind of panoramic, three-dimensional review, and each event seemed to be accompanied by a consciousness of good or evil or with an insight into cause or effect. Not only did I perceive everything from my own viewpoint, but I also knew the thoughts of everyone involved in the event, as if I had their thoughts within me. This meant that I perceived not only what I had done or thought, but even in what way it had influenced others, as if I saw things with all-seeing eyes. And so even your thoughts are apparently not wiped out. And all the time during the review the importance of love was emphasized. Looking back, I cannot say how long this life review and life insight lasted, it may have been long, for every subject came up, but at the same time it seemed just a fraction of a second, because I perceived it all at the same moment. Time and distance seemed not to exist. I was in all places at the same time, and sometimes my attention was drawn to something, and then I would be present there.[65]

Life reviews such as the one described above can include "long forgotten details of their earlier life."[66] If NDEs were hallucinations, we would expect the descriptions to be distorted. But they're not distorted. Rather, they describe real events. As stated by Dr. Long: "The consistent accuracy of life reviews, including the awareness of long-forgotten events and awareness of the thoughts and feelings of others from past interactions, further suggests the reality of NDEs."[67]

"Veridical" out-of-body experiences while clinically dead

But that's not the only place where we see evidence that at least parts of NDEs are real experiences rather than hallucinations. Often NDErs have an "out-of-body experience" (OBE) during which they claim they were floating above their body. And in some of those cases, what they report having seen during the OBE—while clinically dead—matches what actually happened. Such verified OBEs are known as "veridical OBEs." In one study of 96 reported veridical OBEs, 92 percent were shown to be completely accurate.[68] Dr. Long marvels: "The fact that OBErs report seeing and hearing at a time when their physical eyes and ears are not functioning could have profound implications for scientific thinking about consciousness."[69]

Drs. Baruŝs and Mossbridge have examined the data on such cases and conclude: "On the weight of the existing evidence, we think that at least some of the perceptions that occur during NDEs are veridical and that, quite likely, some of them occur during times that, according to the current knowledge of neuroscience, there is insufficient brain activity to support such conscious perception through the usual senses—specifically, a lack of cortical activity."[70]

This is remarkable. So remarkable that it's worth pausing on this to review noteworthy examples of veridical OBEs to get a fuller flavor of how they work.

In his 1975 book, *Life After Life*, Dr. Raymond Moody remarked:

> Several doctors have told me, for example, that they are utterly baffled about how patients with no medical knowledge could describe in such detail and so correctly the procedure used in resuscitation attempts, even though these events took place while the doctors knew the patients involved to be "dead." In several cases, persons have related to me how they amazed their doctors or others with reports of events they had witnessed while out of the body. While she was dying for example, one girl went out of her body and into another

> room in the hospital where she found her older sister crying and saying, "Oh, Kathy, please don't die, please don't die." The older sister was quite baffled when, later, Kathy told her exactly where she had been and what she had been saying, during this time.[71]

Moody reports of another account in which the NDEr claims: "After it was all over, the doctor told me that I had a really bad time, and I said, 'Yeah, I know.' He said, 'Well, how do you know?' and I said, 'I can tell you everything that happened.' He didn't believe me, so I told him the whole story, from the time I stopped breathing until the time I was kind of coming around. He was really shocked to know that I knew everything that had happened."[72]

Similarly, when NDEr Anita Moorjani came out of her coma, she recalls: "I could describe many of the procedures I'd undergone, and I identified the doctors and nurses who'd performed them, to the surprise of everyone around me."[73]

Or consider the famous case of Pam Reynolds, as summarized by the International Association for Near-Death Studies:

> In order to remove a life-threatening aneurysm deep in her brain, Pam Reynolds underwent a rare surgical procedure called "Operation Standstill" in which the blood is drained from the body like oil from a car, stopping all brain, heart and organ activity. The body temperature is lowered to 60 degrees. While fully anesthetized, with sound-emitting earplugs, Pam's ordeal began. Dr. Spetzler, the surgeon, was sawing into her skull when Pam suddenly heard the saw and began to observe the surgical procedure from a vantage point over his shoulder. She also heard what the nurses said to the doctors. Upon returning to consciousness, she was able to accurately describe the unique surgical instrument used and report the statements made by the nurses.[74]

The "dentures" case is also commonly referenced. As reported by a nurse of a coronary care unit:

> During night shift an ambulance brings in a 44-year-old cyanotic, comatose man into the coronary care unit. He was found in coma about 30 minutes before in a meadow. When we go to intubate the patient, he turns out to have dentures

> in his mouth. I remove these upper dentures and put them onto the "crash cart." After about an hour and a half the patient has sufficient heart rhythm and blood pressure, but he is still ventilated and intubated, and he is still comatose. He is transferred to the intensive care unit to continue the necessary artificial respiration. Only after more than a week do I meet again with the patient, who is by now back on the cardiac ward. The moment he sees me he says: "Oh, that nurse knows where my dentures are." I am very surprised. Then he elucidates: "You were there when I was brought into hospital and you took my dentures out of my mouth and put them onto that cart, it had all these bottles on it and there was this sliding drawer underneath, and there you put my teeth." I was especially amazed because I remembered this happening while the man was in deep coma and in the process of CPR. It appeared that the man had seen himself lying in bed, that he had perceived from above how nurses and doctors had been busy with the CPR. He was also able to describe correctly and in detail the small room in which he had been resuscitated as well as the appearance of those present like myself. He is deeply impressed by his experience and says he is no longer afraid of death.[75]

And one more, from a patient who had complications from surgery and then had an NDE with an OBE:

> No, I'd never heard of near-death experiences, and I'd never had any interest in paranormal phenomena or anything of that nature. What happened was that I suddenly became aware of hovering over the foot of the operating table and watching the activity down below around the body of a human being. Soon it dawned on me that this was my own body. So I was hovering over it, above the lamp, which I could see through. I also heard everything that was said: "Hurry up, you bloody bastard" was one of the things I remember them shouting. And even weirder: I didn't just hear them talk, but I could also read the minds of everybody in the room, or so it seemed to me. It was all quite close, I later learned, because it took four and a half minutes to get my heart, which had stopped, going again. As a rule, oxygen deprivation causes brain damage after three or three and a half minutes. I also heard the doctor say that

> he thought I was dead. Later he confirmed saying this, and he was astonished to learn that I'd heard it. I also told them they should mind their language during surgery.[76]

If NDEs are hallucinations, we wouldn't expect NDErs to report facts during their NDEs that can be validated. We would expect delusions instead.

NDEs in cardiac arrest patients—prospective studies

These phenomena baffle mainstream scientists. So some researchers have begun to conduct "prospective cardiac arrest" studies to examine NDEs in a more controlled manner. "Prospective" means that the researchers work with hospitals to collect brain and heart data on patients likely to have NDEs. Rather than finding people who had an NDE in the past, perhaps many years ago, prospective studies allow researchers to interview survivors within days of being resuscitated from cardiac arrest.

Why cardiac arrest? Because the patient's heart is stopped, so blood flow to the brain stops, and there is no electrical firing in the brain (known as a "flat EEG measurement"). Their brains are off.

Dr. van Lommel summarizes what happens during cardiac arrest: "In this state the brain can be compared to a computer that has been disconnected from its power supply, unplugged, and all its circuits disabled. Such a computer cannot function; in such a brain even so-called hallucinations are impossible."[77]

University of Virginia professors Dr. Emily Williams Kelly, Dr. Bruce Greyson, and Dr. Ed Kelly similarly comment: "Cardiac arrest…is a physiologically brutal event. Cerebral functioning shuts down within a few seconds."[78] To be clear: patients in cardiac arrest are clinically dead.[79] Using mainstream materialist assumptions, we would guess that people would not have lucid experiences when their heart and brain are off.

In four separate studies, published between 2001 and 2006, researchers examined 562 cardiac arrest patients at various hospitals. The result: "A number of these patients experienced a period of exceptionally lucid consciousness."[80]

Stop reading for a moment. Reflect on what these researchers stated and how remarkable it is.

We are talking about people in cardiac arrest. Their hearts had stopped. Their brains had no electrical firing. No brain function. They were clinically dead. Doctors running formal prospective studies have the medical records to prove it. And yet some of the surviving patients reported having lucid NDEs while in cardiac arrest! How could the brain have produced these experiences under such brutal conditions? How could any brain-based "hallucination" occur if the brain is not functional?

Dr. van Lommel ran one of these four prospective cardiac arrest studies, which was published in the peer-reviewed medical journal *The Lancet* (2001).[81] The other studies were run by Dr. Greyson (2003), Drs. Sam Parnia and Peter Fenwick (2001), and Dr. Penny Sartori (2006).[82]

In 2014, Parnia et al. released additional prospective cardiac arrest findings in the medical journal *Resuscitation*. They describe a patient who had an NDE and clearly verified events that occurred *during* the time of his cardiac arrest. Medical records validated that the experience happened during cardiac arrest. So one can't argue that the patient somehow had the NDE before or after cardiac arrest. What the patient said he saw during this OBE was indisputably *during* cardiac arrest, when the brain was "off." And yet the person had a lucid, conscious experience.

The patient described hearing an automated voice, saying, "Shock the patient, shock the patient." This was later verified as the noise made by the automated external defibrillator used during his procedure. Based on medical records, the timing of this occurrence and other lucid memories occurred for at least three minutes *during* cardiac arrest.[83]

Dr. Parnia comments on his 2014 finding:

> This is significant, since it has often been assumed that [these] experiences are likely hallucinations or illusions, occurring either before the heart stops or after the heart has been successfully restarted, but not an experience corresponding with "real" events when the heart isn't beating. *In this case, consciousness and awareness appeared to occur during a three-minute period when there was no heartbeat.* This is paradoxical, since the brain typically ceases functioning within 20-30 seconds of the heart stopping and doesn't resume again until the heart has been restarted. Furthermore, the detailed recollections of visual awareness in this case were consistent with verified events[84] [emphasis added].

How do these findings stack up against NDE criticisms?

These findings seem to address popularized criticisms of NDEs, such as the one made by Dr. Sam Harris in his 2014 *New York Times* best-selling book *Waking Up*. He states: "There is generally no way to establish that the NDE occurred while the brain was offline."[85] Yet in the previous example, the NDE apparently *did* occur while the brain was offline. Further, Dr. Harris questioned Dr. Eben Alexander's NDE account. After Dr. Alexander described his NDE in his *New York Times* bestseller *Proof of Heaven*, Dr. Harris wrote: "The very fact that Alexander *remembers* his NDE suggests that the cortical and subcortical [brain] structures necessary for memory formation were active at the time. How else could he recall the experience?"[86] [emphasis in original].

Harris's question is very logical if we adopt the materialist assumption that the brain is responsible for memory and consciousness. Yet, in the aforementioned cardiac arrest studies, we saw that survivors' brains were "off," as confirmed by medical records, and they still had lucid memories. So the findings suggest that the brain alone isn't responsible for memory and consciousness, which negates Harris's criticism.

Another argument I sometimes hear is: "Maybe in these extraordinary cases there is a minuscule amount of lingering brain activity that we simply can't measure due to the limits of technology." If this argument were true, it would still imply the need for a radical shift in neuroscience. The notion that an unmeasurable amount of brain activity is somehow causing the hyper-real accounts and lucid thought processes discussed would be remarkable on its own.

Dr. Parnia addresses this topic: "When you die, there's no blood flow going into your brain. If it goes below a certain level, you can't have electrical activity. *It takes a lot of imagination to think there's somehow a hidden area of your brain that comes into action when everything else isn't working*"[87] [emphasis added].

The blind can see during NDEs

The strangeness of NDEs reaches another level when examining cases of the blind. Some people who have been blind their whole lives report *being able to see* when they have an NDE. Sometimes the reports of what they "saw" during their NDE were verified by people who can see.

As psychiatrist Stanislav Grof, MD, suggests, this phenomenon is yet another counter to the idea that NDEs are mere hallucinations of a dying brain: "There are…reported cases where individuals who were blind because of a medically confirmed organic damage to their optical system could at the time of clinical death see the environment.…Occurrences of this kind… can be subjected to objective verification. They thus represent the most convincing proof that what happens in near-death experiences is more than the hallucinatory phantasmagoria of physiologically impaired brains."[88]

The lead researchers in this area are Kenneth Ring (professor emeritus of psychology at the University of Connecticut) and psychologist Sharon Cooper. In their 2008 book, *Mindsight*, they describe 21 individuals who are blind or severely visually impaired and had NDEs.[89] Ring and Cooper summarize what they found among the blind subjects: "Their narratives, in fact, tend to be indistinguishable from those of sighted persons with respect to the elements that serve to define the classic NDE pattern, such as feelings of great peace and well being…the sense of separation from the physical body, the experience of traveling through a tunnel or dark space, the encounter with the light, the life review, and so forth. In short, the story blind persons tell of their journey into the first stages of death is the common story we have come to associate with these episodes."[90]

One such example is Vicki Umipeg, who suffered optic nerve damage as a premature baby, leaving her completely blind. She remarks: "I've never been able to understand even the concept of light."[91] At age 22, Vicki was involved in a serious car accident and was taken to the hospital. During that time, she had an NDE and recalls being out of her body and "up on the ceiling" watching a male doctor and a woman working on her body. She recalls looking down and recognizing a body and thinking, "I knew it was me…I was quite tall and thin at that point. And I recognized at first that it was a body, but I didn't even know that it was mine initially,"[92] after which she ascended through the ceiling toward a light. She recalls seeing the roof of the hospital, city lights, and buildings. Additionally, she had a life review that she describes as being similar to "seeing a movie, but yet being in it at the same time—and yet I was separate from it.…I saw everything—and I also felt everything that I felt plus everything that everybody else felt."[93]

In Vicki's description of her life review, she notes that she was able to see events of her childhood, whereas when she experienced those events as a blind child, she couldn't see. In her words:

> When I was going into the dining room or into the dormitory generally, of course, I would perceive the things by bumping into them, or by touching them, or whatever. This time I could see them from a distance. It was not like I had to be right on top of them, touching them, or sitting on them or whatever, before I was aware of them....I don't imagine things very well in my mind until I get there. I have a lot of trouble dealing with images of things when I'm not directly there. This time, it was like I didn't have to be right there to be aware of the chairs. I saw the metal chairs that we sat on as children and the round tables in the dining room, and they had plastic table cloths on them. I didn't have to touch the plastic table cloths to be aware of them.[94]

Vicki also entered into an "otherworldly" segment of her NDE, during which she recalls being aware of trees, birds, grass, and flowers. She struggled to describe the flowers when asked by the researchers: "It was different brightness. That's all I know how to describe it as. And different shades....But I don't know. Because I don't know how to relate to color."[95] She claims that what she saw in her NDE has "no similarity, no similarity at all" to what she experiences in dreams. During her dreams she has no visual perception: "No color, no sight of any sort, no shadows, no light, no nothing."[96]

Pretty remarkable given that she has been blind since birth.

The case of Brad Barrows is similarly fascinating. During his NDE, Brad felt himself ascending through the ceiling and then being able to see clouds, slushy snow on the streets, snowbanks that the plows had created, a trolley, and two playgrounds—even though he had been completely blind since birth. He states, "I remember...being able to see quite clearly....I knew that somehow I could sense and literally see everything that was around me."[97] Brad gives a strikingly detailed account of the snow he saw on the streets, which had fallen the day before his NDE: "I think that everything except for the streets were covered with snow, thoroughly. It was a very soft snow. It had not been covered with sleet or freezing rain. It was the type of snow that could blow around anywhere. The streets themselves had been plowed, and you could see the banks on both sides of the streets. I knew they were there. I could see them. The streets were slushy. The snow had fallen when it was almost at the freezing mark, so it was basically slushy. The snow was very soft, kind of wet."[98]

Another striking case is that of Marsha. She was born with severely limited vision, yet she had an NDE during which she described being able to see a black tunnel, a white light, stars, and her body. When she returned from her NDE and felt herself back in her body, she recalls, "I wake up and that's it. I mean, I woke up and I was asking for the light, you know, 'Where's the light?'"[99]

Another man who is blind in one eye and partially blind in the other writes about his NDE: "I came back from my NDE; when I opened my eyes and discovered that the arc of sight was reduced from the previous moment: I HAD SEEN THE PATH TO AND THROUGH THE TUNNEL THROUGH <u>BOTH</u> MY EYES!"[100] [capitalization and emphasis in original].

Or consider an account told by psychiatrist Brian Weiss, MD. This time the story is told from the physician's lens. Dr. Weiss's friend, a prominent cardiologist at Mount Sinai Medical Center in Miami, describes an incident with a patient who underwent cardiac arrest. During the resuscitation procedure, the cardiologist remembers dropping his "distinctive gold pen." The doctor saved the woman, and later she claimed she saw the whole procedure from outside her body. The cardiologist said to her, "You couldn't have. You were unconscious. You were comatose!" She replied, "That was a pretty pen you dropped….It must be valuable." As Dr. Weiss recalls it, she "proceeded to describe the pen, the clothes the doctors and nurses wore, the succession of people who came in and out of the ICU, and what each did—things nobody could have known without having been there."

The cardiologist was still "shaken days later" when he related this to Dr. Weiss. Why? Well, for one, how could she have seen anything while unconscious and comatose? But even more shocking: The woman had been blind for more than five years![101]

Transcendental awareness rather than "seeing"

These people can't see, and yet they are describing visual experiences during NDEs. How can that be? And are the experiences truly "visual"? Ring and Cooper conclude that the "vision" reported is really a form of what they term "transcendental awareness." It's similar to what sighted NDErs describe when they recall omnidirectional, 360-degree awareness.

This would explain why NDErs who are deaf are able to hear what's being said during their NDEs[102] and why those who are color-blind can see "dazzling colors."[103] They aren't actually "hearing" or "seeing," as defined by human language. Rather, they are transcendentally aware, and the best way to describe it using human language is to say they "heard" or "saw."

Ring and Cooper summarize their amazement: "What the blind experience is perhaps in some ways more astonishing even than the claim that they see. Instead, they—like other sighted persons who have had similar episodes—may have transcended the brain-based consciousness altogether and, if that is so, their experiences will of necessity beggar all description or convenient label. For these we need a new language altogether, as we need new theories from a new kind of science even to begin to comprehend them."[104]

Shared-death experiences

In spite of mounting evidence to the contrary, the argument that NDEs are simply hallucinations persists as the leading (unproven) mainstream NDE explanation. The argument is that the NDE is a physiological symptom of a dying brain. However, a phenomenon known as the "shared-death experience" (SDE) further counters the "dying brain" hypothesis.

An SDE is an experience in which a *healthy bystander* has an NDE-like experience. A bystander. The person is not dying. The person is just there with a loved one who is dying. And that bystander has an experience that's like an NDE. So there's no way we can conclude that the bystander is having a "dying brain" hallucination because the person having the experience is not dying. And Dr. Raymond Moody alone has studied hundreds of these cases.[105]

Dr. Eben Alexander sums up the significance of this phenomenon: "Because these shared-death experiences occur in normal, healthy people, they provide powerful evidence against the hypothesis that fundamental elements of near-death experiences—such as a bright light, a tunnel, witnessing departed loved ones, encountering divine begins—are pathophysiological errors of the dying brain."[106]

In one example of an SDE, a healthy woman describes what happened when her husband, Johnny, died in her arms:

> Our whole life sprang up around us and just kind of swallowed up the hospital room…in an instant. There was light all around.…Everything we ever did was there in that light. Plus

> I saw things about Johnny....I saw him doing things before we were married. You might think that some of it might be embarrassing or personal, and it was. But there was no need for privacy, as strange as that might seem. These were things that Johnny did before we were married. Still, I saw him with girls when he was very young. Later I searched for them in his high school yearbook and was able to find them, just based on what I saw during the life review during his death.[107]

That's right—she experienced a "shared" life review.

In another case, a woman describes an SDE in which she lost her mother after trying to perform CPR on her. The woman recalls suddenly lifting out of her body, hovering over it with her now-deceased mother. She recalls: "I looked in the corner of the room and became aware of a breach in the universe that was pouring light like water coming from a broken pipe. Out of that light came people I had known for years, deceased friends of my mother. But there were other people there as well, people I didn't recognize but I assume they were friends of my mother's whom I didn't know....Then the tube closed down in an almost spiral fashion, like a camera lens, and the light was gone."[108]

SDEs can also happen to multiple people simultaneously. In Dr. Raymond Moody's 2010 book, *Glimpses of Eternity*, he explains: "It is possible for a skeptic to easily write off a dying person's death experience when it is shared with only one other person. But a death experience shared with a number of people at the bedside is more difficult to pass off as an individual fantasy."[109]

In fact, Dr. Moody had one such experience himself. When his mother was about to die, he and his siblings stood by her and held hands. Four out of the six of them felt like they were being lifted from the ground. They all reported the light in the room changing—it was like "looking at light in a swimming pool at night."[110] Dr. Moody says, "It was as though the fabric of the universe had torn."[111]

In *Glimpses of Eternity*, Dr. Moody lays out the common events reported in SDEs. No one SDE has reported all seven of these steps, but every SDE reports more than one of these steps:[112]

1. *Change of geometry:* The room changes shape or an alternate reality is experienced.

2. *Mystical light:* Some SDErs report a bright light that emits peace, love, and purity.
3. *Music and musical sounds*: One man described it as "the soft, wild notes of an Aeolian harp."[113]
4. *Out-of-body-experience:* SDErs sometimes report floating out of the body and seeing the scene, almost as a third-party observer.
5. *Co-living a life review:* As summarized by one SDEr: "I was standing in front of what felt like a large screen with my husband who had just died as we watched his life unfold before us. Some of what I saw I had not known before."[114]
6. *Encountering unworldly or "heavenly" realms:* SDErs report accounts similar to what NDErs describe.
7. *Mist at death:* Some SDErs see a mist ("white smoke" or "subtle steam") that is emitted from the dying person.

SDEs haven't been popularized as much as NDEs. But according to Dr. Moody, reports of SDEs are on the rise. He notes that people are often afraid to tell others about their experiences because they fear they will be ridiculed or considered crazy. For example, a financial planner who had an SDE said, "If my business associates heard about this, my name would be mud. Do you think any of them would let me handle their money?"[115]

However, the number of reported SDEs might understate how often they occur. Dr. Moody recalls describing SDEs in lectures, and 5 to 10 percent of the people would raise their hands saying it had happened to them. Dr. Moody claims that "it was only slightly less than the number who raised their hands when asked if they'd had near-death experiences."[116]

If SDEs are occurring with such regularity, why hasn't science explored them more deeply? Further investigation seems like a worthy endeavor. The implications for understanding consciousness are too profound to ignore.

Fear-death experiences

Similar to SDEs, another NDE-like phenomenon is reported in people with ostensibly healthy brains—it is known as a "fear-death experience." These experiences resemble NDEs but occur in people who simply *think* they are about to die. These people experience "acute fear of death" when in a situation with "a seemingly inevitable death."[117] For example, a person might have an NDE-like experience if "almost having a traffic accident or a mountaineering accident."[118] The person escapes unharmed, but still had an NDE-like experience.[119]

What should we make of all this?

NDEs, SDEs, and fear-death experiences certainly challenge conventional models of the brain's relationship to consciousness. Dr. Bernardo Kastrup's whirlpool metaphor is instructive here. If we consider reality to be made up of a stream of consciousness, then our individual experiences are localizations ("whirlpools") of consciousness. Our ordinary, everyday perceptions fall within that localized whirlpool. If the whirlpool were to dissipate and delocalize, other parts of the stream would suddenly become accessible. Perhaps a temporary delocalization process is what happens in the experiences described in this chapter. Perhaps the broader stream of consciousness is no longer obfuscated from our perceptions.[120] We access what was there the whole time (the broader reality of consciousness). It was simply drowned out by the noise of our everyday, localized perceptions.

So these experiences might truly be a tearing in the fabric of the universe—an opening of the filter—rather than mere hallucinations caused by brain physiology. If we drop the materialist paradigm that says that "the brain creates consciousness"—and if we instead contend that "consciousness is fundamental"—then the experiences described in this chapter should be taken seriously—as lessons about the true nature of reality.

Chapter Summary

- NDEs are highly lucid experiences in people who almost died and sometimes were clinically dead. Across cultures and religions, people report similar NDEs (though not identical). There are roughly 15 common features of an NDE, including an out-of-body experience, a life review, visitations from deceased relatives, and overwhelming feelings of love.
- Mainstream materialist science tries to explain NDEs by way of brain mechanisms, commonly arguing that NDEs are hallucinations. But those hypotheses fail to explain key elements of NDEs.

- Sometimes, NDErs see themselves outside of their bodies and are able to verify experiences that occurred during the time of impaired brain function.
- The blind report being able to see during NDEs.
- Even *healthy* bystanders can have NDE-like experiences when a loved one dies (known as a shared-death experience).
- Simply thinking one is about to die can cause an NDE-like experience—known as a fear-death experience—in otherwise healthy individuals.
- If consciousness is the fundamental medium of reality—rather than a product of the brain—then NDEs, SDEs, and fear-death experiences can be viewed as glimpses into the broader reality that is normally hidden from our everyday perceptions.

CHAPTER 10

Communications with the Deceased
Planned and Spontaneous

Physical death does not *entail the end of consciousness, for consciousness is the fabric of all existence*[1] [emphasis in original].

—Bernardo Kastrup, PhD, philosopher

If the brain was the creator of consciousness, then when a person's brain died, their consciousness would cease. Case closed. But if our brains are really TV systems—or antenna receivers—for our minds...then if our brains die...our consciousness hasn't died, i.e., the signal hasn't disappeared because that was not the origin of the signal in the first place. And therefore other people's minds and brains, for example the minds of "mediums," could pick them up. And that's why I've done so much research under single-blind, double-blind and triple-blind conditions, and they all point to the conclusion: that some mediums are real and that the information strongly supports the idea that consciousness survives physical death.[2]

—Dr. Gary Schwartz, professor at the University of Arizona, former professor at Yale University, and Harvard PhD

The evidence, when viewed as a whole, provides more reason for believing in some form of personal postmortem survival than for believing in any alternative view.[3]

—Stephen Braude, PhD, philosopher

> *It seems to me that there is in each of the main areas I have considered a sprinkling of cases which rather forcefully suggest some form of survival. At least—the supposition that a recognizable fragment of the personality of a deceased person may manifest again after his death.*[4]
>
> —Alan Gauld, PhD, former president of the Society for Psychical Research

> *Telepathy and clairvoyance, as we now see, indisputably imply this enlarged conception of the universe as intelligible by man; and so soon as man is steadily conceived as dwelling in his wider range of powers, his survival of death becomes an almost inevitable corollary. With this survival his field of view broadens again. If we once admit discarnate spirits as actors in human affairs, we must expect them to act in some ways with greater scope and freedom than is possible to the incarnate spirits which we already know.*[5]
>
> —Frederic W. H. Myers, one of the founders of the Society for Psychical Research in the late 1800s

NDEs are instructive in that they give hints about what might happen when we die. But let's remember that they are called *near*-death experiences for a reason: The person ultimately lives. When people actually die, what happens to their consciousness? Does it cease to exist, or does it remain in a different form? If it remains, can that consciousness communicate with us?

A mainstream cultural assumption in 2018 is that when the brain dies, the mind dies. However, that is just the popular assumption of choice right now. I cannot see the mind of another person when that person is living. When I walk into a business meeting, I see people who I assume have minds, but I can't *see* their minds. So why do we just assume that the mind dies when the body dies? If we can't see people's minds when they are "alive," isn't it at least possible that the mind continues after bodily death? Maybe we simply can't see a dead person's mind with our eyes, just like we can't see the mind of someone living.

Anecdotal cases and controlled research on this topic suggest that the mind survives bodily death. And the evidence even suggests that "dead" people can communicate with us. These ideas aren't so crazy if consciousness doesn't come from the brain. If consciousness is fundamental, then one's consciousness would remain even if the body dies.

Mediums

Let's start with a definition. A "medium" (sometimes referred to as a "psychic medium") is a person who claims to be able to communicate with people who are dead. I'm not talking about people who had a near-death experience and were resuscitated. I'm talking about actual dead people. In some cases, the medium enters a trance and appears to be overtaken by a foreign consciousness that uses the medium's body and vocal cords to communicate with the living.

The media has popularized mediums in often-preedited TV episodes.[6] Additionally, you've probably seen advertisements for psychics who claim they can speak with your dead relatives. Certainly not all of them are talented, and it's not unreasonable to think that some are outright frauds.

However, there is a solid body of evidence suggesting that *some* mediums are in fact legitimate and that consciousness survives bodily death. Compelling case literature dates back to the 1800s. Stephen Braude, PhD, professor emeritus of philosophy at the University of Maryland, Baltimore County, closely examined the accumulated evidence. In his 2003 book, *Immortal Remains*, he takes a notably conservative and critical approach to the data, but even with that lens he ultimately concludes "that the evidence provides a reasonable basis for believing in personal postmortem survival."[7]

Mrs. Piper

One of the reasons why Dr. Braude makes this statement is due to the talent of a medium named Mrs. Leonora E. Piper (1857–1950), a resident of Boston, Massachusetts. Mrs. Piper was studied extensively by researchers and was known for entering a trance during which she would speak in a voice that was not hers. For example, she would sometimes speak as a French doctor, in a voice that Alan Gauld, PhD, describes as "gruff and male and made use of Frenchisms, and also of slang and swearwords, in a manner quite unlike that of the waking Mrs. Piper." Dr. Gauld notes that her trance was often "accompanied by unpleasant spasmodic movements, grinding of the teeth, etc. There was never the least doubt that the trance-state was in some sense 'genuine'—in it Mrs. Piper could be cut, blistered, pricked and even have a bottle of strong ammonia held under her nose without being disturbed."[8]

Mrs. Piper's results were remarkable. Through the personalities she brought forth while in a trance, she consistently received accurate information about deceased individuals. Either verbally or through writing, she gave accurate messages to "sitters" (the term used to describe living individuals to whom the deceased give messages).

Researchers took precautions to ensure that Mrs. Piper was not cheating. For example, as Dr. Gauld summarizes: "For some weeks Mrs. Piper was shadowed by detectives to ascertain whether she made enquiries into the affairs of possible sitters, or employed agents to do so. She was brought to England where she knew no one and could have had no established agents. During her stay there in the winter of 1889-90, all her sittings were arranged and supervised by leading members of the [Society for Psychical Research]. Sitters were for the most part introduced anonymously, and comprehensive records were kept. And still Mrs. Piper continued to get results."[9]

For example, during her trip to England, one sitter was Sir Oliver Lodge, whose Uncle Robert's twin died 20 years prior to the reading. Lodge handed Mrs. Piper a watch that Robert had given him, which had belonged to his now-deceased twin. Almost immediately, Mrs. Piper reported (while in a trance state) that a deceased man was saying, "This is my watch, and Robert is my brother, and I am here."[10] Robert's deceased twin (through Mrs. Piper) then accurately described specific events in the twins' childhood, such as swimming in a creek as boys and risking drowning, killing a cat, possessing a small rifle, and possessing snakeskin. Details were later confirmed by Robert and his living brother, Frank.[11]

Mrs. Piper even impressed Harvard psychologist William James. In a summary he provided in 1886, he said, "I am persuaded of the medium's honesty, and of the genuineness of her trance; and although at first disposed to think that the 'hits' she made were either lucky coincidence, or the result of knowledge on her part of who the sitter was, and of his or her family affairs, I now believe her to be in possession of a power as yet unexplained."[12]

Mrs. Leonard

Mrs. Gladys Osborne Leonard (1882–1968) is another example of a medium who was studied extensively. She was known for going into a trance and speaking as a foreign personality named Feda who spoke in a high-pitched, childish voice, making "occasional grammatical errors and misunderstandings of words."[13] Part of Mrs. Leonard's notoriety is from her performance on "book tests" in which she would tell a sitter to look at specific pages of a book that Mrs. Leonard had never seen. In one striking case, Feda told an anonymous sitter named Mrs. Talbot that her deceased husband wanted her to look at a message on page 12 or 13 of a particular book at her home.

As Dr. Gauld summarizes:

> Feda said the book was not printed, but had writing in it; was dark in colour; and contained a table of Indo-European, Aryan, Semitic and Arabian languages, whose relationships were shown by a diagram of radiating lines. Mrs. Talbot knew of no such book, and ridiculed the message. However when she eventually looked, she found at the back of a top shelf a shabby black leather notebook of her husband's. Pasted into this book was a folded table of all the languages mentioned; whilst on page 13 was an extract from a book entitled *Post Mortem*. In this case, the message related to a book unknown to the medium and sitter (indeed, so far as could be told, to any living person), but undoubtedly known to the communicator [Feda].[14]

Hafsteinn Bjornsson

Another talented medium was Hafsteinn Bjornsson, an Icelandic man born in 1914. In some instances, he is reported to have spoken languages he had never learned (a phenomenon we will see in chapter 11 as well). For example, when a Danish professor attended one of Bjornsson's séances, Bjornsson began speaking as a deceased individual in the Eskimo language, one he had never learned and which almost no one in Iceland spoke. It turns out that the Danish professor had spent time in Greenland where Eskimo is spoken, knew Eskimo, and knew the person Bjornsson was communicating as.

In another story often referenced by researchers, Bjornsson's consciousness was regularly being overtaken in trance by a man named Runki (later identifying himself as Runolfer Runolfsson), who said, "I am looking for my leg. I want to have my leg." During séances, Runki would exhibit behaviors that were not Bjornsson's. For example, Runki asked for snuff and would make motions with his hands as if he were lifting snuff to his nose to sniff.

In one sitting, a man named Ludvik Gudmundsson joined for the first time, and Runki expressed pleasure in meeting him, saying that his leg was at Gudmundsson's house. Runki gave details of who he was, that he had died while being swept away at sea and lost his leg when dogs and ravens tore his body apart. He also said he was tall. Gudmundsson investigated the matter. He spoke to people in the village near his home and "some of them recalled vaguely that a thigh bone had been 'going around.'" Eventually, Gudmundsson learned that the carpenter who built the house may have put the femur in between two of his walls. Gudmundsson tore down the wall and found a long femur.

University of Iceland professor Erlendur Haraldsson and University of Virginia professor Ian Stevenson investigated this case. They conducted interviews in Iceland with more than 20 people involved, including Bjornsson, sitters who attended Bjornsson's séances, and relatives of the deceased man Runki. Additionally, they found an Icelandic clergyman's records that seemed to confirm that a man named Runolfer Runolfsson had indeed died in the manner described by Runki. Haraldsson and Stevenson carefully studied the strengths and weaknesses of the case, and they concluded in their 1975 article in the *Journal of the American Society for Psychical Research*: "It may be simplest to explain this…as due to Runki's survival after his physical death with retention of many memories and their subsequent communication through the mediumship of Hafsteinn [Bjornsson]."[15]

The totality of the case literature

The Runki case, like virtually every anecdotal case of that nature, is by no means bulletproof. To properly analyze any case, one must examine all of the facts to rule out fraud or other "nonparanormal" explanations. A number of researchers have looked at the case literature painstakingly, including Dr. Braude (as mentioned previously); Dr. Gauld in his 1982 book, *Mediumship and Survival: A Century of Investigations*; and Frederic W. H. Myers in his 1903 book, *Human Personality and Its Survival of Bodily Death*. In Dr. Braude's words, "The totality of survival evidence has a kind of cumulative force, even if individual strands of evidence are less than convincing on their own merits."[16]

Recent controlled studies

In addition to the significant body of historical, anecdotal evidence from mediums like Mrs. Piper, Mrs. Leonard, and Bjornsson, controlled studies have also been conducted. Funding is limited for such studies, as you might imagine. But we do have some data. The leading U.S. research institution today is the Windbridge Research Center, run by Dr. Julie Beischel, who holds a PhD in pharmacology and toxicology from the University of Arizona. She carries the baton from Dr. Gary Schwartz, whose research on mediums is reported in his books *The Afterlife Experiments* (2002) and *The Sacred Promise* (2011).

Dr. Beischel's profile isn't one you might expect for a mediumship researcher. Personal experiences drew her into the field. She initially became interested in mediumship near the completion of her PhD, after her mother committed suicide. At the suggestion of a relative, she saw a medium with the hopes that the medium could communicate with Dr. Beischel's deceased mother. But she was very skeptical. During her session, she jotted down what the medium told her. After later confirming facts with relatives, she found that the medium was 93 percent accurate. As she says, "The reading was very identifying of my mom."[17]

However, the medium seemingly botched part of the reading. She described seeing a man that she claimed was Dr. Beischel's friend as teenager, who died at age 17. Dr. Beischel recalls that the medium claimed "that he drove a restored Mustang or other 'muscle car,' that drinking was involved, that he was aggressive, that he was in my close group of friends, and that he and I joked around a lot."[18] Dr. Beischel had no such person in her life. As far as she knew, this reading was 0 percent accurate.

She later started dating a man named Corey. She did not know Corey at the time of the reading with the medium. When Dr. Beischel mentioned the story of the medium to Corey, he was blown away. The teenager who died in a car accident as described by the medium matched one of *his* best friends in high school. But Dr. Beischel hadn't even met Corey when she'd spoken with the medium! How could the medium have accurately described Corey's deceased best friend—before Dr. Beischel had even met Corey?

This changed Dr. Beischel's life—so much so that she gave up a career in the pharmaceutical industry to pursue controlled research on mediums, applying her scientific mind-set to this understudied area.

Mediumship research is Dr. Beischel's area of focus at the Windbridge Research Center. Her basic premise: Let's see if we can show whether mediums can truly provide information about dead people that they could not have otherwise known, and let's make sure we tightly control the experiments. To do so, she first finds talented mediums after an extensive screening process. Only after they are "certified" can they participate in her studies. The certified mediums are asked in these studies to provide information about a volunteer's (the sitter's) deceased relative (known as a "discarnate"). As Dr. Beischel summarizes: "The medium, the sitter, and the three researchers are all blinded to different pieces of information."[19] In other words, these studies are more-than-double blinded (they are five-way, or quintuple-blinded). And the studies are done *over the phone* so that the medium can't be accused of simply reading physical cues (known as "cold reading").

Here's the catch. The sitter isn't on the phone with the medium; rather, Dr. Beischel is. And all she knows about the discarnate is his or her *first name*. She provides the discarnate's first name to the medium and asks:[20]

- ❍ What did the discarnate look like in his/her physical life? Provide a physical description of the discarnate.
- ❍ Describe the discarnate's personality.
- ❍ What were the discarnate's hobbies or interests? How did she/he spend her/his time?
- ❍ What was the discarnate's cause of death?
- ❍ Does the discarnate have any specific messages for the sitter?
- ❍ Is there anything else you can tell me about the discarnate?

There's no way a medium could know such specific details about a random person's deceased relative under controlled conditions…right?

Wrong, according to Dr. Beischel's studies. The studies show strong statistical results. They suggest that mediums *can* in fact get information about dead people that we can't explain by chance. The results of Dr. Beischel's studies were published in 2007 and 2015 in the peer-reviewed journal *EXPLORE: The Journal of Science and Healing*. Dr. Beischel was then profiled in *New York Times* best-selling author Leslie Kean's 2017 book, *Surviving Death*.

Dr. Beischel summarizes the results and implications of her studies: "These statistically significant…accurate data from a combined total of seventy-four mediumship readings performed under more-than-double-blind conditions that eliminated fraud, experimenter cueing, rater bias, and cold reading show that mediums report accurate and specific information about discarnates and with no sensory feedback. In other words, certain mediums have unexplainable (by current materialist science) abilities to say correct things they shouldn't otherwise know about dead people."[21]

More replications are clearly needed, but the initial results are compelling.

Are mediums actually communicating with dead people?

The above studies suggest that somehow mediums are able to tap into consciousness beyond our everyday realm. Earlier in the book we discussed examples of other psychic phenomena, such as remote viewing and telepathy, in which people are also able to receive information. How do we know that the information mediums receive isn't telepathic information from someone who is still living? How do we know if mediums are actually talking to dead people?

Psychologist Jeff Tarrant, PhD, examined brain activity in mediums and psychics. He tested Windbridge certified medium Laura Lynne Jackson's brain activity and comments: "Laura reports that she sees psychic information in her left visual field, and she sees mediumship information on her right visual field. Actually, this is exactly what we see on these brain images. So this appears to be confirmation of what Laura reports from her own experience."[22]

This finding affirms what mediums report. They notice a difference in how they receive information. For example, one medium said: "A psychic reading is like reading a book...a mediumship reading is like seeing a play." Another noted: "In a mediumship reading, it feels like someone is talking *to* me. With psychic readings, it's information *about* someone" [emphasis in original]. And another said: "It's very different. It's like listening to someone versus looking myself."[23]

So perhaps mediums aren't just telepathically receiving information from living people (which on its own would be remarkable). We might then infer that a medium is actually picking up information from a dead person's consciousness.

Spontaneous after-death communications

Mediumship research suggests that certain individuals are able to communicate with the deceased. But can the deceased communicate with us directly without the help of a talented medium?

The reports in this area are anecdotal, but they seem to occur with regularity. Some research suggests that more than 50 percent of people report experiences of communications from a deceased loved one.[24] Are they imagining it? Does this occur simply because people *want* to believe that their loved ones never really die? Or might some of those encounters be real?

Consider the story of Paul Davids, a Princeton graduate and Hollywood writer and producer (e.g., *Transformers*). His late mentor, Forrest Ackerman ("Forry"), did not believe in life after death, but told Paul and other friends that if he ended up being wrong that he would "drop a line."[25]

When Forry died, strange things began to happen. Davids is now convinced that Forry has been communicating with him and others. In his nearly 500-page book, *An Atheist in Heaven: The Ultimate Evidence for Life After Death?*, Davids provides details of 142 incidents that he feels indicate communications from the now-deceased Forry. Many of these communications even seem consistent with Forry's personality.

A noteworthy example: Davids was at home alone reading through papers. He went to the bathroom, then came back five minutes later. He found that several words were neatly crossed out with wet ink. The specific words crossed out could be construed as a message from Forry. Davids then brought the inkblot to a leading chemist, chairman of the Chemistry Department at Purdue. Chemists spent three years analyzing the ink and were unable to identify the composition. Part of Davids's book is written by John Allison, PhD, a forensic chemist and professor. He explains the chemistry tests in detail and is baffled. Dr. Allison acknowledges his skepticism, but states: "There are amazing things that happen in this world that we need to better understand."[26] Dr. Gary Schwartz writes in the book's preface that at first he was concerned that the story might be a "sham" because it was so unbelievable. But then he writes: "WARNING—THE BOOK YOU ARE ABOUT TO READ IS ABSOLUTELY TRUE" [emphasis in original].[27]

Dr. Schwartz introduced Davids to two mediums, Catherine Yunt and Orit Ish Yemini Tomer. As Davids says, Yunt and Tomer were "carefully kept in the dark. They didn't know who I was and were not told my name. I was simply the 'man with the video camera; who wanted to hear from a deceased friend.'"[28] Apparently the mediums had an impact on Davids. For example, in one instance, Yunt gave Davids such specific and accurate information about Forry that Davids states: "I almost fell out of my chair."[29]

In the book, Davids includes a copy of his signed notarized affidavit, stating: "I swear that I have not invented any of the described phenomena, and that none of these claims are exaggerations and none are the work of imagination or fiction."[30] Furthermore, the book includes a signed letter from Dr. Michael Shermer, executive director of the Skeptics Society, who handwrote: "To Paul, In respect for your honest research and integrity."[31]

Bill and Jody Guggenheim took a different approach. Rather than examining repeated incidents of communications from one deceased person, they looked at a wide range of cases. Over a seven-year period, they collected more than 3,300 firsthand accounts of such after-death communications, summarized in their 1995 book, *Hello from Heaven!*

Their findings show that people experience the deceased in different ways. There are reported accounts of sensing a presence, hearing a voice, feeling a touch, smelling a fragrance, and seeing appearances while awake and in one's dreams.

Here's an example of an account reported:

> Tom and I grew up together. We were next-door neighbors, but I hadn't seen him since he entered the priesthood. I lost complete contact with him and his family after I moved to Texas. One night over ten years later, I woke out of a sound sleep. I saw Tom standing at the bottom of my bed in a Navy uniform! When I saw his uniform, I couldn't believe it because I thought he was a Catholic priest! He said, "Good-bye, Melinda. I'm leaving now." And he disappeared. My husband woke up, and I told him what had happened. But he said I was just dreaming. Three days later, I got a letter from my mother stating that Tom had just been killed in action. I also found out that he had been a chaplain in the Navy![32]

Deathbed visions

Visitations are also reported to occur during "deathbed visions." In a study by University of Virginia psychologist Dr. Emily Williams Kelly, she reports that 41 percent of the dying patients she examined experienced such a vision."[33] However, there have been few scientific studies on the topic.[34]

A deathbed vision is a short experience that occurs in one's dying days, often resembling what is described separately in NDEs. A person may report, "Encounters with deceased loved ones...the sight of a beautiful, unearthly environment and a bright light, or a sense of unconditional love." Reports of deathbed visions typically come from people at the bedside of the dying person. The dying person usually dies shortly after the vision occurs. This is reminiscent of what happens to terminal lucidity patients (referenced in chapter 2) who have unexplained clarity shortly before they die. In some deathbed vision cases, similar to what we saw in shared-death experiences, a bystander at the bedside also perceives what the dying person sees.[35]

So, is life after death real?

If we consider the totality of the evidence—from NDE research, to research on mediums, to anecdotal cases of other after-death communications, we have to seriously consider the possibility that consciousness survives the death of the physical body. And if that is true, we probably need to reconsider what the terms "death" and "life" really mean (more on this in chapter 13).

The notion that consciousness survives bodily death makes no sense under the materialist assumption that the brain produces consciousness. Materialists would argue that once the brain dies, consciousness ceases. But if consciousness is the basis of reality, survival of bodily death not only makes sense, but is expected.

Chapter Summary

❍ "Mediums" are individuals who claim they can talk to the dead. Some do it while in a trance state in which they are apparently overtaken by a foreign consciousness.

❍ The historical case literature, when taken together, suggests that some mediums are real.
  — Mrs. Piper, Mrs. Leonard, and Hafsteinn Bjornsson are examples of mediums who have been studied extensively and appear to exhibit true abilities.

❍ Dr. Julie Beischel of the Windbridge Research Center recently conducted two well-controlled studies on mediums.
  — She found in both studies that mediums were able to obtain information about deceased individuals that they would have no way of knowing by ordinary means.
  — Her findings were published in peer-reviewed journals and were profiled by a *New York Times* best-selling author in 2017.

- There are also reportedly spontaneous encounters with the deceased.
  - Hollywood producer Paul Davids reports 142 incidents in which his deceased friend allegedly communicated with him.
  - Bill and Jody Guggenheim conduced a seven-year study of more than 3,300 individual reports of after-death communications.
  - Dying patients often report "deathbed visions" shortly before death. They claim to see deceased relatives and have a pleasant experience reminiscent of an NDE.

Chapter 11

Lives Beyond This One
Young Children Who Remember Previous Lives

After studying all the cases I have and reviewing the notes of [Dr. Ian Stevenson's] investigations, I have concluded that some young children do appear to possess memories and emotions that come from a deceased individual.[1]

—Jim Tucker, professor of psychiatry and neurobehavioral sciences at the University of Virginia, and director of the Division of Perceptual Studies

Something seems to encourage continuity of personality both within and between lives.[2]

—Dr. Ed Kelly, professor of psychiatry and neurobehavioral sciences at the University of Virginia and Harvard PhD

We should accept reincarnation as the most suitable interpretation for the cases only if we find other interpretations unsatisfactory. This is my situation. I do not believe the correspondences observed between wounds and birth marks or birth defects is a matter of acausal coincidence.... I therefore believe that reincarnation is the best explanation for most of the cases.[3]

—Dr. Ian Stevenson, former professor at the University of Virginia

There are three claims in the ESP field which, in my opinion, deserve serious study: *(1) that by thought alone humans can (barely) affect random number generators in computers; (2) that people under mild sensory deprivation can receive thoughts or images projected at them; and (3)* that young children sometimes report details of a previous life,

> which upon checking turn out to be accurate and which they could not have known about in any other way than reincarnation[4] [emphasis added].
>
> —Astronomer Carl Sagan in his 1996 book, *The Demon-Haunted World*

We've looked at psychic phenomena that suggest consciousness is not localized to the body. We've looked at the strange events that happen around the time of death. And we've looked at evidence that the deceased can communicate with the living. But none of these examples tell us anything about whether our individual consciousness has more than one life experience. Some call that idea "reincarnation."

If we assume that the brain creates consciousness, then reincarnation makes no sense. It's perhaps not surprising that materialists don't buy into it. For example, materialist physicist Lawrence Krauss calls reincarnation "nonsense."[5]

However, if we regard consciousness as fundamental, then reincarnation becomes conceivable. Dr. Kastrup's whirlpool analogy provides a useful framework here. When a whirlpool dissipates, the water simply changes form. It doesn't leave the overall stream. It is "recycled." So if we consider reality to be like a stream of consciousness containing a number of individual localizations (whirlpools), we can imagine that an individual's consciousness is simply recycled within the broader stream. Whirlpools form, then dissipate, then form again. In the same sense, individuals are born, they die, and they are reborn again. Consciousness doesn't die; it simply changes form.

In this chapter we will explore the scientific research suggesting that there might be something to this analogy. As you read, you'll begin to understand why Carl Sagan felt the area deserved attention.

Beliefs about reincarnation

Before diving into the evidence, I should remind you that what is documented here is another example of simply following evidence. Reincarnation, for some, is an emotionally charged subject because of existing beliefs. This book aspires to be detached from preconceptions or wishful thinking. I certainly hadn't heard much about reincarnation before I began my research. It was a completely foreign topic to me that had no basis in reality, as far as I knew. It was an idea that I assumed people made up to comfort themselves about their mortality.

Lead reincarnation researcher Dr. Ian Stevenson of the University of Virginia reminded us: "Critics of the evidence for reincarnation have sometimes pointed to its element of hopefulness with the dismissing suggestion that such evidence as we have derives only from wishful thinking. This objection wrongly assumes that what we desire must be false. We might be more easily persuaded to believe what we wish to believe than the contrary; nevertheless, what we wish to believe may be true. Our inquiry into the truth or falseness of an idea should proceed without regard to whether it fortifies or undermines our wishes."[6]

Children who remember previous lives

Research on children who remember previous lives began around 1960. Then-department head of psychiatry at the University of Virginia (UVA) Medical School, Dr. Ian Stevenson, heard about such children and became intrigued. He then devoted the remainder of his life to this study, examining more than 2,500 cases around the world, until his death in 2007. Dr. Jim Tucker, also a professor at UVA, has continued Dr. Stevenson's research.

Dr. Stevenson's work is highly regarded. Dr. Larry Dossey remarks:

> [Stevenson] reported thousands of cases of children who remembered past lives and whose descriptions of previous existences checked out on investigation.[7]...No one else has researched this area with the scholarship, thoroughness, and dogged devotion to detail as he has. Stevenson combed the planet, from back roads of Burma and the remote villages of India to the largest cities on Earth. He devoted decades to scouring every continent except Antarctica, investigating always the same quarry—children who appear to remember a past life. The scope of his work is breathtakingly universal, and even skeptics are generally awed by the thousands of cases he has amassed. The cases occur in every culture including our own and demonstrate strong internal consistency.[8]

Furthermore, Dr. Stevenson received praise from the well-respected *Journal of the American Medical Association* in 1975: "In regard to reincarnation [Stevenson] has painstakingly and unemotionally collected a detailed series of cases from India, cases in which the evidence is difficult to explain on any other grounds."[9]

What did Dr. Stevenson find?

He found common themes in the cases reported all over the world and in different cultures: A child between the ages of two and five begins to speak emotionally of a past life, including specific events (typically traumatic ones) that are clustered around the end of some previous life.[10] When a child remembers his or her death, the account described is usually violent.[11] Dr. Stevenson stated: "Too often the children are troubled by confusion regarding their identity and this becomes even more severe in those children who, conscious of being in a small body, can remember having been in an adult one, or who remember a life as a member of the opposite sex."[12]

Dr. Stevenson also noted that age plays a role in a child's ability to describe and recall past-life memories:

> I cannot emphasize too strongly that—with some exceptions—a child who is going to remember a previous life has little more than three years in which to communicate his memories of other persons, and he often has less. Before the age of two or three he lacks the vocabulary and verbal skill with which to express what he may wish to communicate. And from the age of about five on, heavy layers of verbal information cover the images in which his memories appear to be mainly conveyed; amnesia for the memories of a previous life sets in and stops further communication of them.[13]

Often children's traits can be linked to the previous lives they remember, such as fears, preferences, interests, and skills.[14] These traits typically bear no resemblance to those of anyone in the child's current family. In some cases, the traits make no sense for a young child, such as desiring certain foods that the family doesn't eat, or desiring "clothes different from those customarily worn by the family members." Stranger than that are cases in which the child has "cravings for addicting substances, such as tobacco, alcohol or other drugs that the previous personality was known to have used."[15]

In some cases, the person allegedly being reincarnated had made a prediction of the next life before his or her death. In other cases, the child has birthmarks, birth defects, or other biological features that align with events of past lives (to be discussed further in the next section).

In a minority of cases, the child exhibits "xenoglossy": speaking a foreign language he or she hasn't been taught.[16]

Where possible, Drs. Stevenson and Tucker have looked for historical facts demonstrating that the person the child remembered matches the child's description. The degree of historical verifiability varies from case to case, but in some cases the accuracy is astounding. In such cases, it is difficult to imagine how a young child could possess such knowledge without access to some broader consciousness.

James 3

One such case is of James Leininger, a young boy in Lafayette, Louisiana.[17] When James was 22 months old, his father took him to a museum, and he showed an affinity for the World War II exhibit. Prior to going to the museum, James had been pointing at planes flying overhead, but he became much more interested after the museum visit. So his parents bought him toy planes and a video of the Blue Angels, the Navy's exhibition team (formed after World War II). James was obsessed and would crash the toy planes into the family's coffee table, denting and scratching the table while saying, "Airplane crash on fire." After his second birthday, he began having nightmares several times a week. He thrashed around the bed with his legs in the air, yelling, "Airplane crash on fire! Little man can't get out!" When awake, he said, "Mama, before I was born, I was a pilot and my airplane got shot in the engine and it crashed in the water, and that's how I died." He told his dad the Japanese shot his plane as part of the Iwo Jima operation, that the plane was a Corsair (a plane not at the museum James had visited), which flew off of a boat called the *Natoma.* He also mentioned that Jack Larsen was there. An additional strange behavior: James was signing his name "James 3."

James's parents were confused, so they investigated some of James's claims. Ultimately, they discovered that James's description matched the historical facts of the life of James Huston Jr. (i.e., James the second), a pilot on *Natoma Bay* who had flown a Corsair and was shot down (in another plane) by the Japanese. Huston was the *only pilot* killed in the Iwo Jima operation, and eyewitnesses reported that the plane was "hit head-on right on the middle of the engine," after which it crashed in the water and quickly sank. Jack Larsen was the pilot of the plane next to James Huston's plane.

The Hollywood extra

Another case is of a four-year-old boy, Ryan, who was born into an Oklahoma family that was traditionally Christian and did not believe in reincarnation.[18] When Ryan played, he would often act as though he was directing imaginary movies by saying, "Action!" When he would see the Hollywood Hills on TV, he would say: "That's my home. That's where I belong…I just can't live in these conditions. My last home was much better."[19] He also talked about having traveled the world and loved Chinatown, saying it had the best food. Ryan claimed that he chose his mother before he was born.

Eventually Ryan started having nightmares, waking up saying he was in Hollywood and his heart exploded. Confused, his mother bought Hollywood books to see if they would trigger any memories. In one book, Ryan saw a photograph of six men from a 1932 movie called *Night After Night*. He said, "Hey, Mama that's George. We did a picture together. And Mama, that guy's me. I found me." Ryan's mother researched and learned that the man Ryan had identified as George was a movie star in the 1930s/1940s named George Raft. However, Ryan's parents could not identify the person who Ryan claimed was "him."

After investigation with the help of Dr. Tucker, they discovered that the man Ryan pointed to was named Marty Martyn, an extra who had no lines in *Night After Night*. Dr. Tucker tracked down Marty's daughter, and she and Ryan then met in person. Ryan's reaction: "Same face, but she didn't wait on me. She changed. Her energy changed."[20]

Many of the claims Ryan had made lined up. For example, Ryan talked of taking girlfriends to the ocean; Marty had taken girlfriends to the ocean and had been married four times. Ryan had remembered an African American maid, and indeed, Marty had one. Ryan mentioned meeting "Senator Five" in New York; Marty's daughter had a picture of Marty with Senator Ives of New York. Ryan said he was a smoker; Marty smoked cigars. Ryan recalled having a nice home and traveling; Marty had a big house with a swimming pool and traveled the world. Ryan talked about liking the food in Chinatown; Marty had enjoyed a Chinese restaurant in Hollywood. Marty died in a hospital room when he was alone, so it is not known whether a heart attack was the ultimate cause of his death, as Ryan's nightmares would have suggested.

As Dr. Tucker summarizes it: "Many of the details Ryan gave did fit the man he pointed to in the picture, who had a much more exciting life than anyone could have guessed a movie extra would have."[21]

How could these children know such detailed facts at such a young age with no evidence of exposure to the details they report?

No wonder Carl Sagan thought this was an area deserving "serious study."

Birthmarks and physical defects

Dr. Stevenson also found links between previous lives and birthmarks and physical defects. His body of work is robust—he wrote a two-volume book entitled *Reincarnation and Biology*, which is more than 2,000 pages long, with dense scientific text and fine print, covering 200 cases (with photographic evidence).

Amazingly, the birthmarks and physical defects he studied correlate to "previous lives" described by the children he examined. It's one thing to read summaries of Dr. Stevenson's work here in this book, but it's another to see the pictures and detail contained in his literature. In an attempt to simply provide some flavor here, I present several examples out of many.

In some cases, birthmarks correspond to wounds verified by a child's memories. For example, Dr. Stevenson described a Turkish boy who remembered a previous life in which he was stabbed through the liver area. In this life, the boy had a "large depressed birthmark, really a small cavity in the skin, over his liver."[22] In another example, a boy from Burma had "a small round birthmark in his right lower abdomen and a much larger birthmark on his right back. These correspond to wounds of entry and exit on the bandit whose life he remembered."[23]

In the strongest cases, medical records verify that the location of a birthmark matches where a trauma occurred in a deceased person. A Lebanese boy recalled a previous life in which he was drinking coffee before leaving for work one day and was shot in the face. The story was verified by an actual shooting that took place. According to medical records related to the shooting, the bullet entered one cheek, damaged the man's tongue, exited through the other cheek, and the man later died in the hospital. The boy, who claims to be the next incarnation of the murdered man, had birthmarks on each check and had difficulty articulating words that required him to elevate his tongue. Dr. Stevenson reported: "I was able to study the hospital record in this case. It showed that the birthmark on [the boy's] left cheek, which was the smaller of the two, corresponded to the wound of entry, and the larger birthmark on the right cheek corresponded to the wound of exit."[24]

In another case, a Turkish boy was believed to be the next incarnation of a recently deceased relative who died after being shot. The bullet did not exit his head, but the pathologist made an incision to extract the bullet. The Turkish boy was born with a birthmark that corresponded with the location of the incision. Dr. Stevenson commented: "Like many other children of these cases, [the boy] showed powerful attitudes of vengefulness toward the man who had shot [him in the previous life]. He once tried to take his father's gun and shoot this person, but was fortunately restrained."[25] The boy came to his parents in their dreams, before he was born, saying he would be the next incarnation of this same deceased relative.

It gets even weirder.

Dr. Stevenson examined "experimental" birthmarks: cases in which a mark was left on the body of a deceased person in the hopes that the mark would show on the person who later reincarnates. In a case in Thailand in 1969, a boy's dead body was marked with charcoal before he was cremated. He had died from drowning. The next boy that the same mother birthed was born with a birthmark near the location of the charcoal marking. Once the boy was able to speak, he began describing details of the life his deceased brother lived. He also had a fear of water.[26]

In another case in Burma, a girl died after unsuccessful open-heart surgery. Her classmates put a mark in red lipstick on the back of her neck before she was buried, in the hopes that the mark would show in the deceased girl's next incarnation. Thirteen months after the girl's death, her sister gave birth to a girl who had a "prominent red birthmark at the back of her neck in the same location where [the deceased girl's] schoolmates marked *her* with lipstick" [emphasis in original]. Dr. Stevenson commented that she also had a birthmark that appeared as a thin line with "diminished pigmentation that ran vertically from her lower chest to her upper abdomen. This corresponded to the surgical incision for the cardiac surgery during which [the girl] had died."[27]

In other cases, more extreme physical deformities can be linked to traumas experienced by the previous life remembered by the child. A Burmese girl was born with birthmarks near her heart and on her head; she was missing the fifth finger on her left hand, and she had "constriction rings" on her legs, the most dramatic of which was on her left thigh.[28] In the disturbing picture provided by Dr. Stevenson in *Where Reincarnation and Biology Intersect* (shown below), her leg looks as though it had been constricted by something like a rope. But her leg is *naturally* shaped that way, without anything constricting it. It is by no means the typical shape of a leg. When the girl was able to speak, she identified herself as a man who had been tortured (fingers cut, tied in ropes). Dr. Stevenson was eventually able to verify this man's identity. There was indeed a person tortured and killed in the precise manner described by the little girl. Distressed by her birthmarks and deformities, the little girl said, "Grandpa. Look at what they did to me. How cruel they were."[29]

How could she have known such specific details? Why would a young child be saying these things? Why did her body reflect such distinctive deformities that matched the way the man had died?

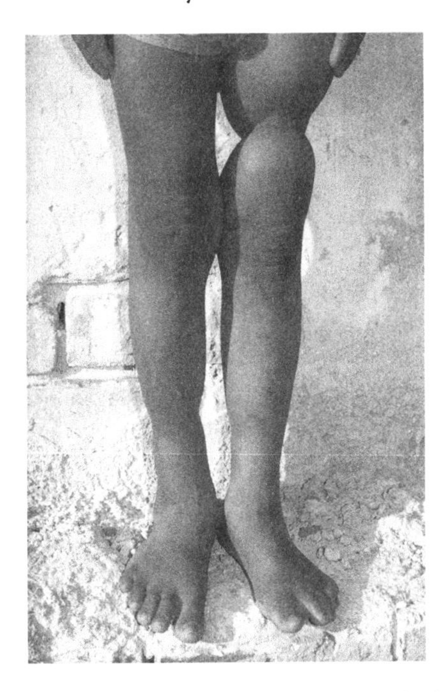

Dr. Ian Stevenson investigated the case of a young girl who had specific, distinctive deformities that match how she described dying in a past life (shown above). In fact, there was a person who died in the precise manner described by the little girl. She described being tied in ropes and tortured in her past life. The shape of her legs seems to match that description. The figure above is from Dr. Stevenson's book *Where Reincarnation and Biology Intersect* (1997).

Dr. Stevenson noted that the girl's mother happened to have walked past the tortured man's dead body when she was two and a half months pregnant. She saw the police handling the situation. She believes the same man came to her in a dream before her daughter was born.[30] Was this little girl the next incarnation of the man who was tortured—even though this man had no biological relationship to the little girl?

Maternal impressions

Dr. Stevenson conceded that in some cases it is not fully clear that birth defects can be tied directly to a previous life. Instead, "maternal impressions" sometimes could have been the cause. Maternal impressions refer to frightening mental images that the mother of a baby has during pregnancy, which result in the baby's having a deformity that closely matches the mental image. Dr. Stevenson referenced an 1890 study at the University of Virginia written by a pediatrician who reviewed 90 maternal-impression cases. In these cases, the mother experienced something particularly frightening during pregnancy. And in 77 percent of the cases, there was "'quite a close correspondence' between the impression upon the mother during pregnancy and her baby's defect."[31]

Dr. Stevenson researched this topic further and focused on 300 cases from around the world, of which he studied 50 in detail. The example that Dr. Stevenson pointed to is disturbing, but makes the point. He recalled the case of a woman whose brother's penis was amputated for medical reasons. While pregnant, "her curiosity impelled her to have a look at the site of her brother's amputation."[32] She then gave birth to a male baby without a penis. Dr. Stevenson researched medical records of the general population and found the odds of a male baby being born with this birth defect is one in 30 million.

Dr. Stevenson's analysis of the cases revealed that maternal impressions most likely impacted the baby when the impression occurred during the first trimester of pregnancy and when the traumatic incident occurred to the mother or someone close to her.[33] Examples such as this caused Dr. Stevenson to question whether every case of physical deformity was induced by reincarnation. Instead, it is possible that some cases could be attributed to maternal impressions (which on its own deserves investigation!).

But Dr. Stevenson noted that sometimes the mother had no knowledge of a deformity during pregnancy. In those cases, the maternal-impressions explanation doesn't hold, and something else (reincarnation?) is needed as an explanation.[34]

Where does this leave us?

The totality of Drs. Stevenson's and Tucker's work points in the following direction, as summarized by Dr. Stevenson: "Some persons have unique attributes that we cannot now explain satisfactorily as due solely to a combination of genetic variation and environmental influences. Reincarnation deserves consideration as a third factor in play."[35] If this is true, the implications are immense for science, medicine, and beyond (as we'll explore in chapter 13).

Under the materialist view that the brain produces consciousness, reincarnation is "nonsense." However, if consciousness is more fundamental than matter and does not arise from brain activity, then the evidence discussed in this chapter is truly plausible.

Chapter Summary

- Drs. Ian Stevenson and Jim Tucker of the University of Virginia have studied more than 2,500 children over 50-plus years who claim to remember previous lives.
- Sometimes the stories these young children tell match historical records of the individuals they claim they were in past lives. In other cases, the children speak foreign languages that they had no way of knowing.
- In some cases, the child has distinctive birthmarks or physical deformities that match the way in which the child describes having died in a previous life. Sometimes, medical records verified the accuracy of the child's claims.

Section V

How Can This Be, and What Does It Mean?

In this section, we will examine how it is possible that science is missing something so big. Additionally, we will examine implications in a variety of domains.

CHAPTER 12

Could Mainstream Science Be So Wrong?

At this point, you may have questions about the material presented thus far. My responses to these hypothetical questions are as follows.

You've covered a lot. Before you give thoughts on whether any of the phenomena are real, can you quickly summarize?

We began by setting the stage and laying the foundation. We established that the origin of consciousness is an open question. Materialists assume that the brain produces consciousness, but have no idea how it happens. An emerging, alternative perspective is that the brain is more like a filtering mechanism for a broader consciousness that exists outside the physical body. The brain is simply a self-localization of consciousness. If we regard consciousness as more fundamental than matter (rather than the reverse), then "paranormal" phenomena are completely normal. To test these ideas, we looked at the relationship between brain activity and the intensity of the conscious experience. If materialism is correct that the brain produces consciousness, we might expect increased brain activity during periods of heightened awareness.

However, we find the opposite. Participants in a study who took psychedelics exhibited reduced brain activity relative to people who were given a placebo. But those who took psychedelics had hyper-real experiences. Similarly, people who have near-death experiences typically have little or no brain functioning and yet they recall lucid memories. These findings make sense if we view the brain as a filter of consciousness: When the brain is less active, the filter is less strong, which allows for a broader spectrum of consciousness to be received. Similarly, in a phenomenon called "terminal lucidity," patients with impaired brains and disorders such as Alzheimer's disease suddenly become completely lucid shortly before dying. Other people with impaired brains (e.g., savants) somehow have extraordinary, seemingly superhuman, mental capabilities.

Elementary animals retain memories even when substantial portions of their brains are removed, rotated, or even replaced. And memories and preferences seem to be transmitted via nonbrain organ donation. All of these examples put into question the view that consciousness comes from the brain.

We then looked at quantum physics, which teaches us that there is proven science that defies common sense—the universe is interconnected ("entanglement"); the act of observing impacts the physical world; matter isn't solid, and we aren't sure what it is; time is relative and may not always work from past to present to future; and space is similarly relative rather than fixed. We also learned that minuscule changes in initial conditions can have a big impact on final outcomes (nonlinear dynamics and chaos theory). On the whole, we learned that we live in a counterintuitive universe, so we shouldn't be surprised that counterintuitive phenomena are real.

With these concepts in our back pockets, we explored the evidence for psychic abilities, all of which suggest that consciousness is not localized to the brain. We began by examining remote viewing, the ability to see at a distance using one's mind alone. We reviewed the government's Stargate Project in which remote viewers used the technique to find a lost, downed plane in an African jungle, among other achievements. Statistical review suggested that remote viewing is real, and even skeptics conceded this. Studies at Princeton further verified the remote viewing results. Remote viewing has even been used by researcher Stephan A. Schwartz to identify archaeological sites. We then looked at evidence for telepathy in multiple areas: the ganzfeld experiment, dreams, the sense of being stared at, telephone telepathy, telepathy among twins, and autistic savants.

Next, we examined ways in which the body appears to know the future before the mind does: the skin, brain, eyes, and heart all react to the future before the future is known. Some people have dreams of the future before the future happens. And precognition might occur before calamities as warnings. We saw that psychic abilities might not be limited to humans. For example, some animals know when their owners are coming home (when the owners mentally *decide* to come home). We also learned that horses might be telepathic, that lost animals can miraculously find their owners great distances away, that they can predict natural disasters, that they know when people will die, and that they can impact the behavior of random number generators with their mental intentions.

In the section's final chapter, we looked at human psychokinesis—the ability to affect matter with the mind. Studies at Princeton University and the Global Consciousness Project show that people's minds have a small, but statistically significant impact on the pattern of randomly generated 0's and 1's. Large-scale psychokinesis has been reported. There are many credible reports that spoons have been bent by the mind alone. Dr. William Tiller, former head of the Material Sciences department at Stanford University, has run studies suggesting that the mind can influence the pH of water and even the development of larvae. Certain energy healers have been studied by scientists. Results suggest that the mind can be used to save cells exposed to radiation. In another case, an energy healer killed cancer cells using his mind.

If consciousness isn't localized to the body, as the above results suggest, then it is conceivable that consciousness survives physical death. We examined the scientific evidence around this idea. NDEs during clinical death (cardiac arrest)—i.e., when the brain is "off"—make us question whether consciousness depends on the brain. The notion that NDEs are hallucinations is difficult to explain if the brain is "off" during the time of highly lucid mental processes. Physiological theories for NDEs struggle to explain all of the elements of NDEs. We also saw that some blind individuals are able to see during their NDEs. Furthermore, shared-death experiences (similar to NDEs) occur in otherwise healthy bystanders. They further suggest that we can't simply claim that NDEs occur as an artifact of a dying brain. Similarly, people have NDE-like experiences when they think they are going to die (fear-death experiences). We then examined after-death communications. Case studies of talented mediums dating back to the 1800s suggest that communications with the deceased do occur. The Windbridge Research Center's recent, controlled studies further suggest that some mediums are real. And in other well-documented cases, the dead spontaneously communicate with the living. Some people have deathbed visions before they die in which deceased relatives visit them. Professors at the University of Virginia have spent more than 50 years studying 2,500-plus children who claim to remember previous lives. The children recall specific details that are in some cases historically verified. Sometimes they speak foreign languages they never learned. Sometimes the children have birthmarks or physical deformities that correspond to wounds from the previous person's death.

Is *any* of this real? Could it really be true that consciousness doesn't come from the brain?

A requirement I hold for myself is that my conclusions should be based on evidence rather than belief. I don't believe in "believing in." Rather, my personal approach is more along the lines of: "Based on the evidence I've seen, I am pointed toward a certain viewpoint that is most likely to be true—even though that viewpoint could change with the introduction of new evidence."

With that preface, I'll point you to two quotations that resonate.

In a 2015 *Scientific American* article, Dr. Michael Shermer, the founder of the Skeptics Society, talked about the need for theories to have a "convergence of evidence from multiple lines of inquiry" rather than "the occasional anomaly."[1]

Jeffrey Mishlove, PhD, summarizes another approach: "Evidence should be considered like a bundle of sticks. Each individual stick might be easily broken, but, when tied together into a bundle, they are as strong as steel."[2]

As I see it, the information presented in this book represents a "convergence of evidence from multiple lines of inquiry" and can be likened to a bundle of sticks rather than just an "occasional anomaly." The evidence points us in the direction of mysterious, nonphysical influences in our reality. In our exploration, we have seen *far* more than "the occasional anomaly." One could attempt to poke holes in any individual area of study, but to disprove *all* of them becomes difficult.

Here's how I think about it. If we look at any individual study or account in this book—which, as you'll recall, is just a summary—we could assume that one of four basic things is happening among the researchers:

1. They are lying.
2. They are delusional (i.e., they are wrong, but mistakenly think they are right).
3. They are using bad scientific or statistical methods (i.e., they are incompetent).
4. They are correct.

Of course, it is possible that some evidence could fall into numbers 1, 2, or 3. But in order to believe that, one would need to provide actual evidence to suggest fraud, delusion, or incompetence. Those are serious accusations. To claim (without evidence) that the investigators must have made it all up is a baseless, unscientific claim. As stated by philosopher Henry Sidgwick in 1882, "We have done all that we can when the critic has nothing left to allege except that the investigator is in the trick. But when he has nothing else left to allege he will allege that."[3]

Unless there is some combination of mass fraud, mass delusion, and mass incompetence across *all* of the various independent scientists over many decades of study—and I have not seen evidence that this occurred in every anomalous example provided—it seems likely that *at least* one phenomenon described in this book falls into category #4.

And if *even one* of the phenomena falls into category #4, then we truly are on the brink of the next scientific revolution—a "Copernican-scale revolution," as Dr. Tiller suggests. I would further contend that if one of the phenomena is real, the odds that others are real increase tremendously because of their interrelatedness.

Others have approached the data similarly. Even back in 1957, the empirical evidence was so great that Professor Hans Eysenck, chairman of the Psychology Department at the University of London, stated: "Unless there is a gigantic conspiracy involving some thirty University departments all over the world, and several hundred highly respected scientists in various fields, many of them originally hostile to the claims of psychical researchers, the only conclusion the unbiased observer can come to must be that there does exist a small number of people who obtain knowledge existing in other people's minds, or in the outer world, by means as yet unknown to science."[4]

Apparently, Eysenck found it hard to believe that such a conspiracy existed. And today we have much more evidence than he had in 1957. I tend to agree with his sentiments.

And recall the quotation from Dr. Jessica Utts, 2016 president of the American Statistical Association, who wrote in a government-commissioned study in 1995:

> Using the standards applied to any other area of science, it is concluded that psychic functioning has been well established.

> The statistical results of the studies examined are far beyond what is expected by chance. Arguments that these results could be due to methodological flaws in the experiments are soundly refuted. Effects of similar magnitude to those found in government-sponsored research...have been replicated at a number of laboratories across the world. Such consistency cannot be readily explained by claims of flaws or fraud.... This is a robust effect that, were it not in such an unusual domain, would no longer be questioned by science as a real phenomenon. No one who has examined all of the data across laboratories, taken as a collective whole, has been able to suggest methodological or statistical problems to explain the ever-increasing and consistent results to date.[5]

And perhaps even more significant is that skeptic Ray Hyman, professor emeritus of psychology at the University of Oregon, was also asked to opine on the same data. Recall his concession that he and Dr. Utts: "Agree on many [other] points. We both agree that the experiments [being assessed] were free of methodological weaknesses that plagued...the early research. We also agree that the...experiments appear to be free of the more obvious and better known flaws that can invalidate the results of parapsychological investigations. We agree that the effect sizes reported... are too large and consistent to be dismissed as statistical flukes."[6]

More recently, in 2011, Patrizio Tressoldi conducted a meta-analysis of studies on psychic phenomena to test whether the "extraordinary claims" have the "extraordinary evidence" required to show that they are real. His conclusion, as published in his article from *Frontiers in Psychology*: "If results analysed with both frequentist and Bayesian statistical approaches from more than 200 studies conducted by different researchers with more than 6000 participants in total and three different experimental protocols are not considered 'extraordinary,' or at least 'sufficient' to suggest that the human mind may have quantum-like properties, what standards can possibly apply?"[7]

Dr. Ed Kelly sums up the situation: "Indeed, we predict with high confidence that future generations of historians, sociologists, and philosophers of science will make a good living one day trying to explain why it took so long for scientists in general to accept [these phenomena]."[8]

In light of all this, my conclusion is that at least some of the topics discussed in this book are likely to be real and that all of them "deserve serious study" (to quote Carl Sagan). That means real attention from the mainstream. No more sweeping these concepts under the rug; no more ignoring the anomalies because they are inconvenient or because someone doesn't want them to be true; no more fear of ridicule for questioning mainstream paradigms; but rather, true, honest exploration to uncover what's real and what's not—through an examination of evidence and data.

Would more data help make the case?

Certainly, additional research is needed, and more data can't hurt. But one could argue that in some cases we already have enough evidence. Instead of proving that the phenomena are real, we might want to be focusing on understanding how they work.

Along these lines, Dr. Utts stated in her 1995 government-commissioned report: "It is recommended that future experiments focus on understanding how this phenomenon works, and on how to make it as useful as possible. There is little benefit to continuing experiments designed to offer proof, since there is little more to be offered to anyone who does not accept the current collection of data."[9]

Similarly, Dr. Robert Jahn, former dean of engineering at Princeton University, said of his nearly 30 years of results from the PEAR lab: "If people don't believe us after all the results we've produced, then they never will."[10]

How could there be so much data that I never knew about? Why isn't this information talked about more often?

The topics are taboo. In today's Western society, there is a cultural bias against them. Some scientists are even afraid to talk about them.

Psychiatrist Brian Weiss, MD, summarizes the situation well: "It is only the reluctance to tell others about psychic occurrences that makes them seem rare. And the more highly trained are the most reluctant to share."[11] He continues: "I understood why these highly trained professionals remained in the closet. I was one of them. We could not deny our own experiences and senses. Yet our training was in many ways diametrically opposite to the information, experiences, and beliefs we had accumulated. So we remained quiet."[12]

Psychologist Dr. Imants Barušs and cognitive neuroscientist Dr. Julia Mossbridge see the same issue (repeating from chapter 1 because it is so significant): "As a result of studying anomalous phenomena or challenging materialism, scientists may have been ridiculed for doing their work, been prohibited from supervising student theses, been unable to obtain funding from traditional funding sources, been unable to get papers published in mainstream journals, had their teachings censored, been barred from promotions, and been threatened with removal of tenured positions. Students have reported being afraid to be associated with research into anomalous phenomena for fear of jeopardizing their academic careers. Other students have reported explicit reprisals for questioning materialism, and so on."[13]

Brenda Dunne of Princeton's PEAR lab echoes these sentiments: "We submitted our data for review to very good journals…but no one would review it. We have been very open with our data. But how do you get peer review when you don't have peers?"[14]

Dunne and her colleagues published an article in 2007 that explicitly makes this point: "The…[RNG study on psychokinesis]…was originally submitted to various segments of the *Physical Review* spectrum of journals in the hope of engaging more members of the physics community in similar research efforts. It was rejected, without any technical reviews, over a series of editorial appeals, on the ideological grounds that it was an 'inappropriate' topic for that scholarly venue. It was subsequently dismissed a priori by the editorial board of *Foundations of Physics*. Ultimately, it was published intact by [the *Journal of Scientific Exploration*], and has been widely referenced since its appearance."[15]

When dealing in this controversial domain, even being a Nobel Prize winner doesn't seem to help. For example, Nobel laureate Brian Josephson states:

> My transition into believing that mind has to be taken seriously as an entity in its own right proved also to be a transition into an environment that was hostile where previously it had been very supportive. The scientific community has its own belief systems that it is dangerous to challenge.…Being a Nobel Laureate protects one from the worst pressures, but not from the curiosities such as this letter relating to a conference to which I had previously been given an invitation and even been asked how long I wished to speak:

> *It has come to my attention that one of your principal research interests is the paranormal.…In my view, it would not be appropriate for someone with such research interests to attend a scientific conference.*

> I learned from subsequent correspondences that it was feared that my very presence at the meeting might damage the career prospects of students who attended, even if I did not touch on the paranormal in my talk.…More seriously, my interest in such matters seems to have led to the harassment of students working with me, even in regard to projects not related to the paranormal.…My original assumption that scientists, being intelligent people, would have the ability to view experimental evidence and theoretical arguments objectively has been severely challenged.[16]

This dynamic is not new, however. Researcher William Crookes reported facing similar resistance to his work on paranormal phenomena, and this was back in the 1800s![17]

Why are so many articles on these phenomena (such as Wikipedia articles) so negative?

A number of organizations are openly hostile to claims of the paranormal. One has to wonder if their sentiments have played a role on Wikipedia and elsewhere.

One such organization is the Committee for Skeptical Inquiry (CSI). CSI's Lee Nisbet, who holds a PhD in philosophy, said of paranormal phenomena: "We feel it is the duty of the scientific community to show that these beliefs are utterly screwball." One of CSI's former co-chairs of its "Committee for the Scientific Investigation of Claims of the Paranormal" resigned because he felt that "they sought to debunk rather than scientifically examine."[18]

Dr. Rupert Sheldrake has been a target of criticism from skeptics and speculates that a group called "Guerilla Skepticism" is contributing to negativity around the paranormal on Wikipedia. As Dr. Sheldrake states in his blog post *Wikipedia Under Threat*:

> Wikipedia is a wonderful invention. But precisely because it's so trusted and convenient, people with their own agendas keep

trying to take it over. Editing wars are common.…Everyone knows that there are opposing views on politics and religion, and many people recognise a biased account when they see it. But in the realm of science, things are different. Most people have no scientific expertise and believe that science is objective. Their trust is now being abused systematically by a highly motivated group of activists called Guerrilla Skepticism on Wikipedia.

Skepticism is a normal, healthy attitude of doubt. Unfortunately it can also be used as a weapon to attack opponents. In scientific and medical contexts, organized skepticism is a crusade to propagate scientific materialism.…Most materialists believe that the mind is nothing more than the physical activity of the brain, psychic phenomena are illusory, and complementary and alternative medical systems are fraudulent, or at best produce placebo effects.…Several advocacy organizations promote this materialist ideology in the media and in educational institutions. The largest and best-funded is the Committee for Skeptical Inquiry (CSI), which publishes *The Skeptical Inquirer* magazine. The Guerrilla Skeptics have carried the crusading zeal of organized skepticism into the realm of Wikipedia, and use it as a soapbox to propagate their beliefs.

This summer…a commando squad of skeptics captured the Wikipedia page about me. They have occupied and controlled it ever since, rewriting my biography with as much negative bias as possible, to the point of defamation.…The Guerrilla Skeptics are well trained, highly motivated, have an ideological agenda, and operate in teams, contrary to Wikipedia rules. The mastermind behind this organization is Susan Gerbik. She explains how her teams work in a training video. She now has over 90 guerrillas operating in 17 different languages. The teams are coordinated through secret Facebook pages. They check the credentials of new recruits to avoid infiltration. Their aim is to control information, and Ms. Gerbik glories in the power that she and her warriors wield. They have already seized control of many Wikipedia pages, deleted entries on subjects they disapprove of, and boosted the biographies of atheists.

> As the Guerrilla Skeptics have demonstrated, Wikipedia can easily be subverted by determined groups of activists, despite its well-intentioned policies and mediation procedures. Perhaps one solution would be for experienced editors to visit the talk pages of sites where editing wars are taking place, rather like UN Peacekeeping Forces, and try to re-establish a neutral point of view. But this would not help in cases where there are no editors to oppose the Guerrilla Skeptics, or where they have been silenced.
>
> If nothing is done, Wikipedia will lose its credibility, and its financial backers will withdraw their support. I hope the noble aims of Wikipedia will prevail.[19]

If what Dr. Sheldrake describes is truly happening, then we might (in part) understand why the reality of the "paranormal" has struggled to gain momentum. The casual researcher doesn't have time to dig into the details. If you Google a topic and the first thing that comes up is a Wikipedia article saying it's fraudulent, you might stop your search then and there.

Why do you think so many scientists resist or reject the ideas discussed in this book?

This book exposes paradoxes and anomalies. As Dr. Dean Radin comments, paradoxes and anomalies have a way of evoking resistance from those holding onto conventionally held beliefs:

> Paradoxes are extremely important because they point out logical contradictions in assumptions. The first cousins of paradoxes are anomalies, those unexplained oddities that crop up now and again in science. Like paradoxes, anomalies are useful for revealing possible gaps in prevailing theories. Sometimes the gaps and contradictions are resolved peacefully and the old theories are shown to accommodate the oddities after all. But that is not always the case, so paradoxes and anomalies are not much liked by scientists who have built their careers on conventional theories. Anomalies present annoying challenges to established ways of thinking and because theories tend to take on a life of their own, no theory is going to lie down and die without putting up a strenuous fight.[20]

Additionally, maybe some scientists simply prefer to avoid phenomena they don't understand. Dr. Ian Stevenson references a telling quote from a University of Virginia pediatrician who examined controversial maternal impressions cases in 1890 (discussed in chapter 11). After becoming aware of growing skepticism around him, the researcher said: "Thinking men came to doubt the truth of those things which they could not understand."[21]

The evidence provided in this book implies that *many* brilliant scientists are wrong in their theories. How could that be?

Yes, this book implies that they are very wrong about certain things, but that shouldn't take away from their brilliance in other areas.

If we were to look back at the leading scientists of several hundred years ago, we could find reasons to poke at their lack of sophistication and call them "clueless" relative to what we now know. For example, in 1772 the father of modern chemistry, Antoine Lavoisier, and his fellow academics examined reports of "stones falling from the sky." They concluded, "Stones cannot fall from the sky because there are no stones in the sky!"[22] Those "stones" were later discovered to be meteorites. Science then began to accept that stones can fall from the sky. Lavoisier was a pioneer in science in some ways, but clearly he was not right about everything.

I wonder what society 200 years in the future will say about the 2018 mainstream scientific community. Will they laugh at how primitive we were? Will they scoff at the notion that we were so ignorant to assume that "the brain creates consciousness"?

Also, we should acknowledge that the topics discussed in this book are multidisciplinary. Think about how many different areas of science have been covered: physics, neuroscience, cardiology, biology, chemistry, etc. Herein lies a serious problem in science: scientists specialize in *their* areas of expertise and are often uncomfortable or unfamiliar with subject matter outside of those areas. Scientists can indeed be very brilliant in their specific areas of expertise, but might know very little about other areas. That can lead to errors or gaps when trying to build unifying theories.

For example, some prominent physicists rarely, if ever, talk about consciousness. Most physicists think that consciousness has no interaction with the physical world, so they view it as a topic for psychology and philosophy. Consciousness is not their area of expertise.

Consider physicist Neil deGrasse Tyson's *New York Times* bestseller *Death by Black Hole: And Other Cosmic Quandaries*—the index of that book does not even include the word "consciousness." Recall that he comments, "I wonder whether there really is no such thing as consciousness at all."[23] The same goes for physicist Brian Greene's books *The Hidden Reality* and *The Fabric of the Cosmos*. Recall that Dr. Stephen Hawking said, "I get uneasy when people, especially theoretical physicists, talk about consciousness,"[24] further proclaiming "philosophy is dead."[25]

The list could go on and on. These men are recognized as some of the greatest minds in the world, and yet they're leaving out a potentially fundamental piece of the puzzle: consciousness.

By the same token, neuroscientists, psychologists, and biologists don't often account for quantum mechanics. Instead, they tend to operate under classical, Newtonian paradigms that only give an approximation for reality. Ignoring the quantum reality can lead to significant mistakes.

It is therefore possible that the lack of interdisciplinary integration has led to significant oversights in science.

Finally, many prominent scientists simply aren't taking the time to look at the data. For example, physicist Lawrence Krauss states: "I don't have the time or inclination to investigate something that is highly likely to be wrong."[26] So perhaps the mainstream scientific belief in materialism keeps scientists from researching the topics discussed in this book. They don't want to spend their time on something they deem impossible.

If the ideas discussed in this book are real, we would need to radically alter science. What's your view on that?

Isn't that what science is all about: refining our theories when we learn of new data? We know how little we know. We know, for example, that 96 percent of the universe is made of hypothetical "dark matter" and "dark energy." We should *expect* that our theories will change, given how little we know. Therefore, we should be open to the possibility of new ideas. That's not to say we should accept every theory we come across, but we should be open to scientifically exploring new ideas.

I'm *not* suggesting that if materialism is proven wrong (or incomplete), that we should just throw it out. We can thank materialism for many technological and medical advances from which we now benefit. The alternative theories discussed in this book simply suggest that materialism is a special case of a broader picture of reality. So we would need to recontextualize, rather than redo, all that we know.

Let's assume that consciousness isn't produced by the brain. Let's assume that "somehow" the brain accesses consciousness from outside the body. You don't explain a mechanism for how this could happen. You tell us about strange things that occur with no explanation of *how* they occur. How does the brain interact with consciousness?

The short answer is: I don't know.

But there is no requirement that one must know *how* a phenomenon works in order to accept that it *does* exist. Dr. Julie Beischel and her colleagues make this point in their 2015 journal article on mediums. The authors explain that mediums appear to obtain nonchance information about the deceased, but they cannot explain how mediums do it.

Beischel et al. remind us that there are many areas of science for which we do not know the cause, but which we accept as being real: Causes are "currently unknown or not fully understood for numerous (1) ubiquitous human experiences (e.g., yawning, dreaming, and blushing); (2) diseases and conditions (e.g., multiple sclerosis, lupus, rheumatoid arthritis, Parkinson's disease, eczema, psoriasis, glaucoma, and fibromyalgia); and (3) medications (e.g., certain drugs that treat Parkinson's [pramipexole], cancer [procarbazine], malaria [halofantrine], and epilepsy [levetiracetam]; the antibiotics clofazimine and pentamidine; and many psychotropic drugs (e.g., [lithium]) which continue to exist, be experienced, be widely prescribed, and be worthy of scientific study even in the absence of a known mechanism."[27]

Dr. Larry Dossey further comments: "In science, we often know *that* something works before we have a clue about *how* it works.... Explanations often come later"[28] [emphasis in original].

There is no question that further study is needed to understand *how* it all works. But that won't happen until the mainstream decides that these areas are worthy of study.

Some emerging theories around consciousness are developing, however. One theory comes from prominent Oxford mathematical physicist Roger Penrose and University of Arizona anesthesiologist Dr. Stuart Hameroff. In the 1990s they initially proposed, and still support, a "microtubules" theory. The two represent a unique combination of skill sets across disciplines: Penrose from the mathematics/physics lens, and Hameroff from the psychology/medicine lens. As they summarize it, their microtubules theory says: "That consciousness depends on biologically 'orchestrated' quantum computations in collections of microtubules within brain neurons [and] that these quantum computations correlate with and regulate neuronal activity."[29]

Furthermore, Dr. Hameroff posits: "Let's say the heart stops beating, the blood stops flowing; the microtubules lose their quantum state. But the quantum information which is in the microtubules isn't destroyed, it can't be destroyed, it just distributes and dissipates to the universe at large. If the patient is resuscitated, revived, this quantum information can go back into the microtubules and the patient says 'I had a near-death experience'.... If they're not revived, and the patient dies, then it's possible that this quantum information can exist outside the body, perhaps indefinitely."[30]

The fact that Roger Penrose's name is connected with the microtubules theory is particularly significant. He is a world-renowned mathematician and physicist who has collaborated (and coauthored books) with Stephen Hawking. Penrose's work is credited for piquing Hawking's interest in black holes and general relativity.[31] However, Penrose's foray into consciousness caused a rift between him and Hawking, as profiled in the May 2017 *Nautilus* article, *Roger Penrose on Why Consciousness Does Not Compute: the emperor of physics defends his controversial theory of mind.*[32]

Would more funding help?

I think so. But not everyone agrees with that stance.

For example, California Institute of Technology physicist Sean Carroll writes in a 2008 blog post: "I would put the probability that some sort of [psychic] phenomenon will turn out to be real at something (substantially) less than a billion to one. We can compare this to the well-established success of particle physics and quantum field theory. The total budget for high-energy physics worldwide is probably a few billion dollars per year. So I would be very happy to support research into [psychic phenomena] at the level of a few dollars per year. Heck, I'd even be willing to go as high as *twenty* dollars per year, just to be safe"[33] [emphasis in original].

In light of Dr. Carroll's comments, it's perhaps not surprising to hear that these phenomena are not well-funded. As Dr. Gary Schwartz says about his research on anomalies of consciousness: "Even in the best of economic times, conventional funding sources—such as the National Science Foundation or the National Institutes of Health (both of which have funded my mainstream research in the past)—are not open to supporting this challenging and controversial research."[34]

Dr. Dean Radin similarly states that this field "threatens the very core of assumptions of science, and it is not easy raising funds to challenge a powerful status quo."[35]

What if large foundations such as the Bill & Melinda Gates Foundation donated even a fraction of their billions of dollars to consciousness-related studies? Wouldn't it be worthwhile for us to more fully explore the nature of our existence? The implications could transform society. Now seems like the time for philanthropic organizations to recognize that we are on the brink of the next scientific revolution. It won't happen without the appropriate funding.

What role can younger generations play in furthering the exploration of these topics?

Nobel Prize-winning physicist Max Planck noted: "A new scientific truth does not triumph by convincing its opponents and making them see the light, but rather because its opponents eventually die."[36]

As crude as it sounds, Planck makes an astute observation. Over time, some of the scientists most opposed to the ideas discussed in this book will no longer be living. They have good reason to want to hold on to their theories—by accepting a new paradigm they would have to admit that they were wrong. Some people are unwilling to take such an ego hit.

And what we are dealing with here are challenges to ideas regarded as *fact*. So the task is even more difficult. As stated by astrophysicist Bernard Haisch: "Modern Western science regards consciousness as an epiphenomenon that cannot be anything but a by-product of the neurology and biochemistry of the brain. . . . While this perspective is viewed within modern science as a fact, it is in reality far stronger than a mere fact: it is a dogma. Facts can be overturned by evidence, whereas dogma is impervious to evidence."[37]

It is incumbent upon younger generations to further the exploration of these topics as the "old guard" loses its grip on science.

Chapter 13

What Are the Implications for Everyday Life?

A host of implications are discussed in this chapter. They are based on my premise that one or more of the "anomalous" phenomena in this book are, in fact, real. And given this premise, I conclude that the materialist models of "Matter is fundamental" and "The brain creates consciousness" are wrong. Based on the data presently available, I view the more likely framework to be that consciousness is the basis of existence (as discussed in the preface). The implications I discuss here are based on this assumption.

While much of the book thus far has relied on hard science, some of the concepts discussed in this final chapter tend to apply logical inference and extrapolation. In some instances, the concepts are inherently abstract and difficult to capture with language. In other cases, the concepts are simply difficult for the limited human mind to comprehend. But we'll explore anyway.

Why do some people have psychic abilities while others do not?

Many participants in the discussed ESP studies were normal people, and psychic effects were shown—even if the effects were small and required statistics to see them. The studies suggest that all humans have these abilities, even if the abilities are subtle.

Some people are more naturally skilled than others. In his 2006 book, *Entangled Minds*, Dr. Dean Radin likens psychic abilities to high-jumping.[1] People have different abilities, and the most talented high jumpers can jump really, really high. But they're all jumping, even if they aren't very good at it.

A small number of psychics are "superstars" and have strong abilities that are somewhat reliable. We discussed a number of them in the context of the U.S. government's Stargate Project on remote viewing (e.g., Joe McMoneagle, Ingo Swann, Uri Geller, etc.).

One has to wonder if superstar psychics' brains are naturally configured in a way that allows for enhanced psychic functioning. The study of savants' brains might also be instructive here. More research is needed.

Can I become a superstar psychic? And how do meditation, yoga, hypnosis, and sensory deprivation fit into the picture?

Many who claim to have psychic abilities go into a meditative trance that allows them to calm the mind and quiet the chatter in their heads, so they can "receive" or "access" information. This is likely what former president Jimmy Carter was describing in regard to the remote viewer who went into a trance before locating the downed plane in the African jungle.

Russell Targ, who worked at the Stanford Research Institute with many of the best remote viewers in the United States, explains that when he guides remote viewers: "My first job is to help the viewer to silence the ongoing mental chatter—mental noise, or the 'monkey mind.'"[2]

Dr. Radin examined the topic of meditation and psychic abilities in his 2013 book, *Supernormal: Science, Yoga, and the Evidence for Extraordinary Psychic Abilities*. He summarizes his analysis of the available scientific literature: "We've seen evidence suggesting that advanced meditation may be associated with improved [psychic] performance."[3]

Hypnosis is similar—a person enters a hyper-relaxed state in which thinking is reduced. There are reports of people under hypnosis who spontaneously begin speaking languages they've never learned.[4] Perhaps their "antennae" are better able to "receive" information while their ordinary mental chatter has been quieted.

The practice of yoga may also be effective because it similarly calms the mind. By focusing on bodily movements, the mind naturally quiets.

And sensory-deprivation flotation pods might serve a similar function. The pods are filled with concentrated salt water such that a person naturally floats. The pods typically have no noise and no light, so a person in the pod effortlessly floats in an antigravity environment, with essentially no external stimuli for the brain to process. The environment encourages a meditative state. It is perhaps not surprising, then, that one of the benefits of floating is to achieve clarity and enhance creativity.[5] In the context of what we just discussed, this makes sense: the environment encourages a quieting of the mind—leaving the "antenna" more open to receive information, whereas in a normal environment, the brain is distracted by the need to process a host of external information.

What is the "third eye"?

There exist theories that psychic functioning involves the use of one's "third eye," more formally known as the pineal gland. As Dr. Diane Powell states: "Evidence does support the idea that the pineal gland plays a role in creating states of consciousness that are conducive to receiving psychic information."[6] Additional research is needed to better understand the role of the pineal gland.

How do these ideas relate to Elon Musk's new brain-centric startup (Neuralink), which seeks to build "a wizard hat for the brain"? [7]

In March 2017, visionary businessman Elon Musk announced the formation of a brain-computer interface company called Neuralink, which, as reported by *The Verge*, "is centered on creating devices that can be implanted in the human brain, with the eventual purpose of helping human beings merge with software and keep pace with advancements in artificial intelligence. These enhancements could improve memory or allow for more direct interfacing with computing devices."[8]

While little information on Neuralink is available, Musk has provided some clues as to what he plans to do. For example, he describes one of his goals to repair damaged brains:

> The first use of the technology will be to repair brain injuries as a result of stroke or cutting out a cancer lesion, where somebody's fundamentally lost a certain cognitive element. It could help with people who are quadriplegics or paraplegics by providing a neural shunt from the motor cortex down to where the muscles are activated. It can help with people who, as they get older, have memory problems and can't remember the names

> of their kids, through memory enhancement, which could allow them to function well to a much later time in life—the medically advantageous elements of this for dealing with mental disablement of one kind or another, which of course happens to all of us when we get old enough, are very significant.[9]

These aims are noble, but if Musk truly wants to build a "wizard hat," he will need to depart from the current paradigms of materialist neuroscience. He will need to explore the possibility that consciousness is not localized to the brain. If Musk can better understand how the brain relates to consciousness, maybe Neuralink will be able to unlock our "inner wizard" by artificially inducing brain states that enable on-demand psychic abilities.

Can you imagine being able to press a button that puts your brain into a meditative state, allowing you to remote view something miles away? Or pressing a button to read someone's mind? Or being able to impact matter with your mind whenever you want?

Ethics would, of course, need to be considered. You probably wouldn't want a stranger to remote view you while you're taking a shower, or for someone to know your thoughts without your permission. And you wouldn't want someone to be able to physically harm you using his or her mind.

But before we get carried away, let's remember how little we know about these topics. Our understanding of the brain is limited, and the brain's relationship to psychic abilities is even less developed. Neuralink cofounder and University of California, San Francisco professor emeritus Philip Sabes acknowledges our current limitations: "If it were a prerequisite to understand the brain in order to interact with the brain in a substantive way, we'd have trouble."[10]

What are the implications for artificial intelligence?

Our limited understanding of the brain may similarly inhibit our ability to understand the potential for artificial intelligence. Some scientists such as Google executive Ray Kurzweil worry that artificial intelligence poses a threat to humanity. In 2011 he stated: "Within a quarter century, nonbiological intelligence [i.e., robots] will match the range and subtlety of human intelligence. It will then soar past it because of the continuing acceleration of information-based technologies, as well as the ability of machines to instantly share their knowledge.... Nonbiological intelligence will have access to its own design and will be able to improve itself in an increasingly rapid redesign cycle. We'll get to a point where technical progress will be so fast that unenhanced human intelligence will be unable to follow it."[11]

There is no doubt that artificial intelligence is becoming more sophisticated. But can the "range and subtlety" of human intelligence be replicated by a computer?

The materialist view is that consciousness is produced by a material brain. So if we can create a brain-like machine, we should be able to create "conscious" artificial intelligence. In other words, we should be able to create machines that have feelings, thoughts, and a sense of experiencing their existences—a sense of "I." And if these machines became intelligent enough, maybe they could take over the world.

That approach is overly simplistic if we adopt the lens established in this book. In order to create truly "conscious" artificial intelligence, we would need to better understand how consciousness is received, filtered, and localized by the brain. Maybe we can't make conscious machines. Maybe we can only make machines capable of doing computational tasks—but which don't have feelings.

These are big ideas and hot topics in the technology community. Consciousness and its relationship to the brain need to be better understood before we can draw conclusions. The abilities, limitations, and threats of artificial intelligence may ultimately depend on our understanding of where consciousness originates.

Can I use psychic abilities to make money?

Some people claim they have. For example, after leaving the Stanford Research Institute in 1982, Russell Targ formed Delphi Associates, which aimed to make money in markets by using psychic abilities. As Targ summarizes it, his group "psychically forecasted the changes in the price of silver successfully nine times in nine weeks, making $120,000, which was a lot of money at that time. As a trader on the floor of the Commodity Exchange told *NOVA*, which documented our exploits, 'Doing anything nine times in a row in this volatile market is impossible!'"[12]

Ethical questions obviously arise when hearing stories such as this. They're fascinating, nonetheless.

Can psychic abilities be used to develop warning systems?

In Dr. Rupert Sheldrake's study of animals' psychic abilities, he notes that animals seem to know in advance when earthquakes, tsunamis, avalanches, storms, and even air raids are coming.[13] A possible explanation is that animals are using their innate psychic abilities.

He suggests that they can be used to forewarn us of natural disasters, stating: "Imagine what could happen in California and other parts of the Western world if, instead of ignoring the warnings given by animals, people took them seriously. Through the media, millions of pet owners could be informed about the kinds of behavior their pets and other animals might show if an earthquake was imminent."[14]

You might be surprised to know that China uses such systems. While the system hasn't been perfect, there have been successes as reported by Dr. Sheldrake: "They have had some spectacular failures, most notably the unpredicted Tangshan earthquake of 1976, in which at least 24,000 people died. But they have continued to make successful predictions. For example, in 1995 they warned local authorities in Yunnan Province one day before a major earthquake struck."[15]

Yanong Pan, a professor at Chaohu College, Anhul Province, was a scientist who has been involved in these studies. He provided an explanation of China's warning system to Dr. Sheldrake, stating:

> Before the earthquake, animals will have different features or feelings. The most important thing is to keep watching them in case something happens. Horses, donkeys, cows, and goats don't want to go back to their stalls. Mice flee from their homes in groups. The chickens fly onto trees, and pigs try to destroy their sties. Ducks and geese are frightened to go into the water, dogs bark in sorrow and pain. Hibernating snakes wake up unusually early. Swallows, pigeons, and other birds fly away. Rabbits have their ears straight up and bounce around, knocking into things. Fish feel they are threatened and are terrified; they stay near the surface of the water without moving. Watch carefully, everyone in each family; observe what's going on, draw your conclusion, and if you are sure tell the government as soon as possible.[16]

Can psychic abilities be used for security purposes, both national and personal?

The reason for the government's 20-plus-year Stargate Project was to use remote viewing for national security. So the government clearly saw value in at least exploring it.

There are also reports that law enforcement uses psychic abilities. Remote viewing researcher Stephan A. Schwartz reports: "Nationwide,...about 100 police departments openly work with Remote Viewers on a regular basis. The other people in law enforcement who consult viewers (there are no statistics on how many) are off the record about it: An officer quietly enlists the assistance on his or her own."[17]

A high-profile example is that of Patricia Hearst's kidnapping (she is the daughter of newspaper magnate William Randolph Hearst). In 1974, retired police commissioner Pat Price used his psychic abilities to track down Hearst's kidnappers.

The Berkeley police contacted the Stanford Research Institute, asking for psychic assistance on the matter. Price was a remote viewer in the program at Stanford and was able to help. His initial psychic impression was that the kidnappers did not want money, and instead, the kidnapping was political in nature. The police showed Price hundreds of unlabeled mug shots, and he picked out three, all of which were part of the group they later learned had kidnapped Hearst. Price noted that one of the men "recently had his teeth pulled out at the dentist without anesthesia, relying instead upon self-hypnosis."[18]

Two days later, the kidnappers contacted the police, and, as Price predicted, they claimed that they didn't want money; instead, they wanted food for the poor. Eventually, police determined that the three men Price picked out were indeed three of the men in this group. Police even confirmed Price's story about the kidnapper's dental incident.[19] Price also used remote viewing to locate the kidnappers' car, which was 50 miles away.[20]

What are the implications for "group" intentions?

We saw in chapter 8 that random number generators (RNGs) around the world can behave nonrandomly when many people are focused on the same thing at the same time. That begs the question: If we collectively *decide* to focus on the same thing at the same time as a large group, can we have an impact on the physical world? Can we do this to change the world in a desired direction?

Researcher Bryan Williams looked at this question in 2013 by examining RNG data from The Global Consciousness Project. He focused his analysis on instances of "global harmony." In other words, the events he selected for his analysis were ones in which there was a collective intention to positively impact society. He found that during 110 such events between 1998 and 2012, RNGs behaved nonrandomly. His interpretation is that "these ephemeral moments when we are 'just wishing'...are effective in the world."[21] The effects are small, but they seem to exist nonetheless.

Dr. Radin gives another perspective on this topic that is worth considering:

> If you have a large pond and you want to get all of the water out of the pond, you could take a rock and throw it into the pond and a little bit of water would be pushed out as a result. So you might think, well, I'll just get 1,000 friends and we'll all throw our rocks into the pond and it will make all the water spill out. Well, that probably won't happen. The reason is that unless you throw the rocks in at exactly the same time and at exactly the right configuration, you won't end up with one giant wave. Instead, you will end up with just a very disturbed surface of a pond because all the waves are interfering with each other. By the same token, if you have one person thinking about peace, that probably does affect the world, a little bit. But if you have 100 people thinking the same thing, if they are not all in exact alignment with each other, then you won't get 100 times more intention. You may end up with zero intention because the waves cancel each other out.[22]

If our minds can impact the physical world, what are the implications for scientific experimentation?

When we talk about a "controlled" scientific study, we assume we are controlling or fixing all variables. In this process, we tend to assume that the mind of the experimenter is completely independent from the experiment. This assumption comes from materialism, which says the brain produces consciousness and as a by-product of the brain, consciousness has no impact on the world itself.

As we discovered in chapter 8 on psychokinesis, the mind may have an ability to impact the physical world. If consciousness is indeed more fundamental than matter, then of course the mind could have an effect on physical events.

This topic is an important one to consider when we evaluate experimental results. If a skeptical experimenter were running a study with the hopes of showing no effect, could that experimenter unknowingly impact the results with his mind? We need to at least be open to the idea that this could be true. Similarly, an experimenter who *wants* an experiment to show results might be able to influence the outcome. And we can't forget about the participants themselves—how might their states of consciousness impact the results?

Science does not currently have the answers, but the scientific community will need to seriously consider these implications—and reconsider empirical results—if we depart from materialism.

Beyond scientific experimentation, are there even broader implications for our understanding of science, technology, and the universe?

If consciousness is more than just a by-product of the brain, then contemporary science is missing something huge. Who knows what kinds of advances we might see if scientists and technologists would alter their methods and theories accordingly? It's hard to project just how significant the advances could be if we have been so fundamentally wrong.

Furthermore, this new understanding of consciousness might help us reframe how we think about the universe. Modern physics struggles to reconcile quantum theory with general relativity: The theories work well independently, but they do not work well when applied together. The inability to devise a unified theory is a thorn in physicists' sides that just won't go away. But what if introducing consciousness into the picture can help?

Some scientists feel that consciousness does indeed help us get closer to a unified theory. For example, cosmologist Dr. Jude Currivan proposes a consciousness-centric view of the universe in her 2017 book, *The Cosmic Hologram*. The crux of Dr. Currivan's premise—and supported by wide-ranging evidence across many fields of research and all scales of existence—is that "everything that manifests in the physical world emerges from deeper and ordered levels of nonphysical and in-formed reality."[23] This perspective regards the universe as existing and evolving as a unified entity—a reality fundamentally made of consciousness, in the form of digitized information. And moreover, that its appearance is simply a holographic projection of its holographic boundary. Dr. Currivan further provides a novel perspective on the "incompatible" quantum and relativity theories—viewing them not as antagonistic, but as complementary expressions of information manifesting as the energy-matter and space-time of our interconnected, holographic reality.

Along these lines, one has to wonder if the incorporation of consciousness into mainstream physics might also help us understand the mysterious "dark energy" and "dark matter" that make up roughly 96 percent of the universe.

While much is still to be understood, what seems clear is that inclusion of consciousness and the significance of information needs to be brought into scientific focus.

What are the implications for more day-to-day issues, like mental health? Can psychedelics help?

Recently, growing attention has been paid to the potential health benefits of psychedelics for people with mental disorders. For example, Johns Hopkins University and New York University have shown impressive results in their studies. In a study on psilocybin (the active psychedelic in "magic" mushrooms), 80 percent of terminally ill cancer patients reported experiencing less anxiety and depression about the prospect of dying six months after being treated with the substance. Two-thirds of them described their experience on the substance as one of their top five most meaningful life experiences.[24]

The notion that psychedelics can be therapeutic is not new, however. Indigenous Amazonians, for example, have used the psychedelic ayahuasca as a healing agent for centuries or more. Ayahuasca is now becoming popular in the United States.[25]

Why might psychedelics be therapeutic? If we consider the brain to be a filter—a "reducing valve"—of consciousness, and if we view psychedelics to partially unlock the filter by reducing brain activity (as discussed in chapter 2), then perhaps psychedelics naturally enable a mind-set shift. Rather than being stuck in a limited reality of anxiety, depression, PTSD, addiction, etc., perhaps getting a glimpse of the broader reality causes a dramatic shift in perspective that traditional "talk therapy" might take years to achieve. This glimpse is seen similarly in other transcendental or mystical experiences such as NDEs and enlightenment or awakening experiences (more on this coming up).

One has to wonder if psychedelics will prove to be more effective than traditional mental health drugs. Further exploration is needed.

What are the implications for medicine and personal health, more broadly?

The idea of integrating "mental intentions" into existing health practices—such as the studies discussed on energy healing—seems worthy of investigation. The current paradigms of Western medicine do not regularly account for the idea that the mind can impact matter. Dr. Bruce Lipton, author of *The Biology of Belief*, notes the predicament that modern medicine faces: "Medical doctors are caught in an intellectual rock and a hard place.... Their healing abilities are hobbled by an archaic medical education founded on a Newtonian, matter-only universe. Unfortunately, that philosophy went out of vogue seventy-five years ago, when physicists officially adopted quantum mechanics."[26]

Dr. Lipton and others, such as Dawson Church, PhD, in his book *The Genie in Your Genes*, emphasize the importance of our thoughts and emotions on our genes. This emerging field, known as epigenetics, "has shaken the foundations of biology and medicine to their core because it reveals that we are not victims but masters of our genes."[27] In epigenetics, gene expression is controlled "from outside the DNA": signals from the environment, including our mental states, "turn genes on and off."[28] Therefore, as Dr. Church reminds us, there are studies showing that a positive attitude can decrease one's risk of cardiovascular disease relative to those who have a more pessimistic outlook.[29] Similarly, stress-inducing situations can cause the release of hormones that negatively impact health. They can kill brain cells and take away the body's resources for cell repair.[30]

Consciousness studies combined with fields such as epigenetics suggest that doctors shouldn't limit their focus to physical causes and treatments of illness. They should also incorporate nonphysical, mental aspects. This suggestion would mark a radical shift in Western medicine. These principles are not traditionally taught to doctors in Western medical schools.

As we saw in chapter 9, Anita Moorjani's mental state shifted during and after her NDE, which caused her terminal cancer of four years to miraculously disappear. Her doctors were stunned and could not explain how or why it happened. As Moorjani puts it: "I understood that my body is only a reflection of my internal state. If my inner self were aware of its greatness and connection with All-that-is, my body would soon reflect that and heal rapidly."[31]

Finally, the reincarnation research conducted by Drs. Ian Stevenson and Jim Tucker raise questions that medicine currently cannot answer. A baseline assumption in mainstream medicine is that physical traits and illnesses are caused by some combination of genetics and environment. However, in cases where children remembered past lives, the children "inherited" memories, preferences, birthmarks, and physical deformities that seem to have no relation to genetics or environment. So there is the potential for an unknown "third factor" that needs to be considered beyond genetics and environment. If a "third factor" does in fact exist, then we need to rethink medicine in a big way.

If consciousness survives the death of the physical body as implied in chapters 9 through 11, how would that change one's view of death (and life)?

Psychic phenomena suggest that consciousness is not localized to the physical body. Research on NDEs, communications with the deceased, and children who remember previous lives takes this idea a step further and suggests that consciousness even survives bodily death. As stated by Sir David Hawkins, MD, PhD: "Consciousness does not depend on physicality but exists independently of it."[32]

Adopting this perspective can radically change one's view of death. Our society tends to assume when the brain and body die, the mind dies with it. But rather than viewing death as an end of life, death could alternatively be considered more like a "transition" from this physical world into some other state of being. If we adopt Dr. Kastrup's whirlpool analogy, which compares reality to a stream of water (consciousness) where each of us is a localized whirlpool, then when the whirlpool delocalizes, it simply flows back into the stream. It doesn't "die." The localization changes form. But it is still just made of water (consciousness). The water itself doesn't disappear from the stream.

NDEs might provide a glimpse into what the transition state is like. Recall that many NDErs are clinically dead—their brains are "off"—and yet they have lucid, often blissful, experiences. One may wonder, then, if NDEs provide a literal preview of what happens to one's consciousness when the physical body dies.

One NDEr summarizes: "When we got to the light, the totality of life was love and happiness. There was nothing else. And it was intense. Very intense and endless in scope."[33] Another NDEr describes: "The sound of that music I cannot possibly describe with words because it cannot be heard with that clarity in this world! The colors were out of this world—so deep, so luminous, so beautiful!"[34] And another recalls: "It seemed so much more real than anything I had ever experienced in my entire life."[35]

People who had NDEs were *near* death, but they didn't actually die. So by looking at NDEs we are simply inferring what might happen at death, though we cannot prove it. If one were to take the many existing NDE accounts as literal descriptions of what happens when the physical body dies, then death should not be feared. However, further study is needed before we can draw conclusions. For example, a minority of NDEs are described as being frightening.[36] How does that fit into our picture of what happens after physical death?

But accounts like engineer Dr. Alan Hugenot's are ultimately comforting. He says that as a result of his transformative NDE: "I have no fear of dying. While I do fear being injured or broken, I simply have no fear of death itself. I know that death is only a transition, and I am fully aware that our consciousness survives this change."[37]

Dr. Elisabeth Kübler-Ross, a preeminent end-of-life researcher, took a similar stance. Years of studying dying patients led her to conclude: "Death is simply a shedding of the physical body like the butterfly shedding its cocoon. It is no different from taking off a suit of clothes one no longer needs. It is a transition to a higher state of consciousness where you continue to perceive, to understand, to laugh, and to be able to grow."[38]

This idea could dramatically impact how we treat the dying, how we grieve for lost loved ones, and how we live life generally. Many people fear death because it is viewed as an "end." If it's not an end—and if it's as blissful as most NDErs report—then perhaps it shouldn't be feared. And maybe we could live happier lives without being burdened by the dread of the end of physical life.

Does the survival of consciousness have implications for whether life has deeper meaning?

Panoramic life reviews described across many NDErs are instructive here. They suggest that a primary reason for living is to learn and to evolve consciousness. In the life review, the NDEr reviews his life from his perspective *and* from the perspective of those he affected in his life, all while judging himself—as if to learn to do better next time—and ultimately working toward a state of unconditional love. There might in fact be a "next time" if one accepts the reincarnation evidence discussed in chapter 11. This notion makes the physical body seem like a temporary conduit or vessel for a broader consciousness. Each individual's body is then the lens through which the broader consciousness has a particular experience in a physical form. And when the physical experience is over (what we call "death"), consciousness can enter another physical vehicle for a different learning experience. If we truly are entangled as "one mind," then the learning experienced by each individual's consciousness is simply a piece of the broader learning and evolution of a universal, collective, "one" consciousness.

Dr. Ervin Laszlo, a philosopher of science and systems theorist twice nominated for the Nobel Peace Prize, concludes his 2016 book, *What is Reality?*, with a succinct summary of the above ideas. He states:

> We are here to evolve the consciousness of the cosmos by evolving our consciousness. We can pursue this task throughout the cycle of our existence. During the incarnate phases, when our consciousness appears to reside in our body, we can evolve our consciousness by fostering its capacity to enter in the domains where nonlocal intuitions and experiences of oneness and unconditional love appear.... When the discarnate phase ends, our evolved consciousness encounters optimum conditions for reincarnation to earthly existence. In the new incarnate phase further opportunities are given for experiencing and for learning, and for evolving to ever-higher forms of consciousness. The cycle repeats. The evolved consciousness ascends to higher and higher Planes of existence.[39]

The above ideas are in stark contrast to the materialist view. I can describe the materialist view easily because I was a materialist for most of my life. Materialism teaches that consciousness arises from our brain. When our brain dies, our consciousness dies. It's like shutting off a computer. Everything that the person had experienced in life had no real meaning beyond what that person made of it. Once you die, none of the artificially created meaning is relevant because your brain is shut off and therefore your consciousness no longer exists. You can try to create meaning while alive, but ultimately whatever meaning you created is fabricated and fleeting. This is what a strict interpretation of materialism implies, as bleak as it sounds.

This book suggests that materialism is wrong. So the above materialist explanation of life's meaning likewise seems wrong. If consciousness is fundamental, then life has to be viewed in a more meaningful light since some part of us doesn't actually die.

How to define that meaning is another issue. One has to wonder if there are limits to what our human mind can know and comprehend. The above theories are attempts at piecing together a story from a few data points.

The perspective you describe sounds extremely comforting, but do you think it's just a rationalization?

As a materialist, I thought comforting theories of existence had to be wrong. I viewed them as rationalizations. To me they were nothing more than methods of hiding from the inevitability of death. However, in hindsight, I realize I was making a logical error. I was reasoning that because a theory of existence was comforting, it also had to be false. I overly discounted the possibility that a theory of existence could be both comforting *and* true. After having studied the evidence, I now have a very different perspective. I think a comforting picture of existence is likely to be true.

If consciousness isn't just a product of the brain, and if it survives the death of the physical body—beyond space and time—then how does it fit into our picture of the universe?

The double-slit experiment described in chapter 3 suggests that the observer—some would say consciousness—plays a special role in our physical reality. Recall that the observer "collapses the wave function"—until the observer "observes," the particle behaves like a wave of possibilities. But when the observer observes, the wave behaves differently. It behaves like a particle instead of behaving like a wave. And as we saw in

Dr. Radin's potentially revolutionary studies, there seems to be evidence that consciousness indeed affects the wave function.

An extrapolation of these findings could lead one to reason that consciousness is *creating* the physical world we experience. As stated by cardiologist Dr. Pim van Lommel: "Some prominent quantum physicists, including Eugene Wigner, Brian Josephson, and John Wheeler, as well as mathematician John von Neumann support the radical interpretation that observation itself literally creates physical reality, a position that regards consciousness as more fundamental than matter."[40]

Recall that the father of quantum physics, Max Planck, declared in 1931: "I regard consciousness as fundamental. I regard matter as derivative from consciousness."[41]

Like Planck, many others have concluded that consciousness is fundamental in the universe. Consciousness is primary, not matter. It's as if consciousness is creating the material world we see. So to say that the brain produces consciousness is completely backwards. It's the other way around. Consciousness produces the brain!

For example, Dr. Jim Tucker of the University of Virginia states: "Consciousness does not exist because the physical world does; the physical world exists because consciousness does."[42] He adds, "Consciousness is independent of the physical world and even the creator of the physical world."[43]

Likewise, neurosurgeon Dr. Eben Alexander states: "Consciousness is not only very real—it's actually *more real* than the rest of physical existence, and most likely the basis of it all"[44] [emphasis in original].

Stem cell biologist Robert Lanza, MD, and physicist Bob Berman postulate a similar idea, calling it "Biocentrism," which contends: "If there were no observers, the cosmos wouldn't merely look like nothing, which is stating the obvious. No, more than that, it wouldn't exist in any way."[45] Stanford physicist Andrei Linde similarly states: "The universe and the observer exist as a pair. I cannot imagine a consistent theory of the universe that ignores consciousness. I do not know any sense in which I could claim that the universe is here in the absence of observers."[46]

Oxford mathematical physicist Roger Penrose and University of Arizona anesthesiologist Dr. Stuart Hameroff further declare: "We conclude that consciousness plays an intrinsic role in the universe."[47]

Johns Hopkins physicist Richard Conn Henry states that "the Universe is entirely mental," reminding us: "There have been serious attempts to preserve a material world—but they produce no new physics, and serve only to preserve an illusion."[48] (He is referring to the illusion that a physical world exists outside of consciousness.)

And physicist Norman Friedman sums it up: "According to my view of quantum theory, matter does not produce consciousness. It is the other way around. Consciousness produces material events. This is done by 'choosing' from a latent field of probable events."[49]

These concepts take us back to where we started in the preface. The images below illustrate the chain of reality suggested by materialism versus the "consciousness is primary" alternative.

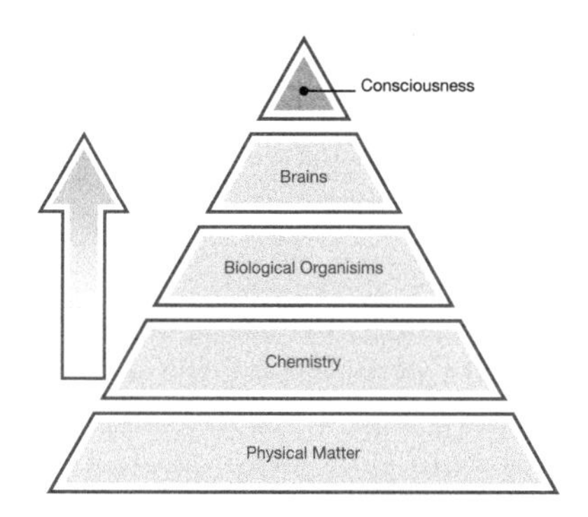

Figure A
Materialism
(consciousness comes from matter)

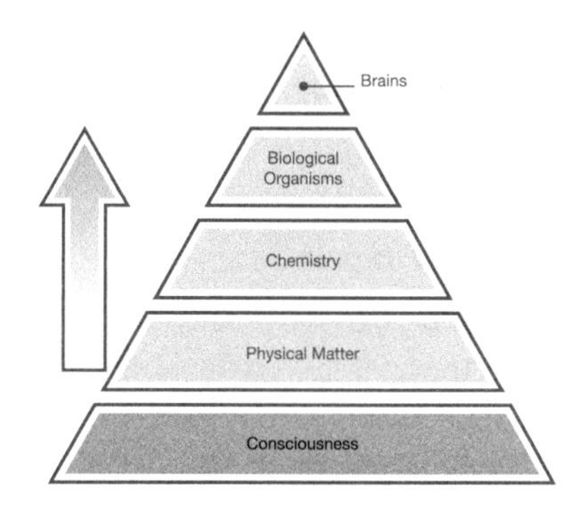

Figure B
An alternative perspective in which consciousness is primary

Under the view in Figure B, which is what is suggested by the evidence in this book, we can understand why Nobel Prize-winning physicist Steven Weinberg stated: "Matter thus loses its central role in physics."[50]

And as phrased in a 2015 compilation of scientific articles written by physicists Roger Penrose, Henry Stapp, and Menas Kafatos; and a host of other scientists: "Consciousness became the universe."[51]

If consciousness is primary, quantum physics and "anomalies" of consciousness start to make sense. They are only anomalies if we assume matter is primary (materialism).

And recall from this book's preface that the current paradigm of materialism (Figure A) is an unprovable belief system. Einstein even called it a religion.[52] We cannot prove that anything exists outside of consciousness, let alone that something (matter) preceded and caused it.

Philosopher Rupert Spira summarizes the conundrum: "No one has ever found, or ever could find, anything outside consciousness."[53] Furthermore, he offers an exercise to demonstrate this idea: "Try now with your attention to leave the field of consciousness in which all experience appears, in the same way that a child might lie in bed wondering how far space goes and what, if anything, might lie beyond it. See that attention never leaves the field of consciousness. All experience takes place in and is known by consciousness, and as experience is all-that-is or ever could be known, we cannot legitimately claim the existence of anything outside of consciousness. To do so would require a leap of faith." [54]

Materialism, as illustrated by Figure A, not only fails to account for the phenomena described in this book, but it is also unprovable.

So are you advocating "panpsychism"?

No, not as panpsychism is often defined. The distinction here is subtle, but important. Panpsychism is sometimes used to address the hard problem of consciousness. The hard problem is that we don't know how physical matter could give rise to a nonphysical consciousness. The panpsychist view asks, "What if consciousness is an inherent property of matter"? In other words, every unit of matter, even a tiny particle, *has* some amount of consciousness. A system's level of consciousness will vary depending on the way in which its matter is configured.

This perspective is a veiled form of materialism, however. It asserts that matter *has* consciousness. So panpsychism starts with matter and says consciousness comes *from* it.[55] By contrast, I am putting forth a framework that makes consciousness the basis of reality. Matter is an experience within consciousness. Using philosophical lingo, the metaphysical picture of reality I'm advocating is sometimes called "monistic idealism": "the notion that all reality is grounded in a transpersonal form of consciousness."[56]

What does this imply about what it means to be human?

From the perspective of materialism, each of us is a body that *has a* consciousness. However, the evidence presented in this book suggests to me that this thinking is upside down. Instead, it seems more likely that we are a consciousness first and foremost that is having the experience of a physical world *through* a body. That is a very different outlook than the materialist perspective. We'll explore notions of identity in further detail shortly.

If that's true, then why didn't I know this about myself? Have I somehow forgotten?

This is really a question about the nature of memory. While memory can be reliable for some things, let's remember how much of our lives we do not remember. For example, I have no memory of being an infant. I don't even remember details of more recent events. If you asked me what I was doing at a specific time in the recent past, say, at 10:07 p.m. on December 8, 2014, I wouldn't have an answer. So should I assume that I didn't exist at that time? No. It simply means that I am unable to recall exactly what happened. The same thing occurs with dreams. We don't remember every detail of every dream we have. So all of this suggests that we have a form of amnesia, whether we regularly acknowledge it or not.

We have to wonder: What else might have happened that we don't remember? Is it conceivable that our true identity is part of some broader consciousness that we have simply forgotten?

Recall from chapter 11 that some young children remember previous lives and then seemingly forget them as they age. Why are they able to remember previous lives in the first place, and why do some eventually forget? Why do some children remember past lives while others do not? Science doesn't have answers yet. However, topics around memory are critical to developing an understanding of our existence and deserve further investigation. In the meantime, we should be mindful of how limited our memories truly are.

The concepts about consciousness you are discussing sound a lot like what mystical traditions have been saying for millennia. Is that accurate?

Fritjof Capra wrote an entire book on this topic, called *The Tao of Physics*, which explores the ties between modern physics and beliefs of mystical traditions such as Hinduism, Buddhism, Zen, and other Eastern religions. Mystical traditions of Western religions—such as Gnosticism

(Christianity), Kabbalah (Judaism), and Sufism (Islam)—echo these ideas as well.

So the notion that "consciousness is primary" is not at all new. It may sound new to some of us in 2018 because of the materialist bias in contemporary academia. But it is not new.

Just to give a flavor of the types of beliefs held by mystics across cultures for ages, consider this quotation from mystic and nonscientist Sri Nisargadatta Maharaj: "Consciousness itself is the source of everything."[57] That sounds a lot like what science is now pointing to.

Let's assume that consciousness is, in fact, the source of everything. Then where does consciousness come from?

This is a natural question, but it is perhaps misguided. It is predicated on the common-sense belief of the limited human mind that something must *cause* something else. As we've learned, our ordinary experience can lead us astray when trying to understand reality beyond approximation.

Let's take a simple example. I hit a tennis ball with a tennis racket, and the ball goes over the net. From a local, narrow perspective of watching the tennis racket hitting the ball, it seems like it was the racket that caused the ball to go over the net.

But let's look at a level deeper. Where did both the tennis racket and ball come from? We can trace the chain of events and find that the number of things that needed to happen in order for the racket to hit the ball is astronomical: The Bang 13.8 billion years ago, leading ultimately to the evolution of life on Earth billions of years later; the evolution of humanity to get to the point where humans could manufacture tennis rackets and tennis balls; I needed to be born in order to hold the racket that hit the tennis ball, and I was only born because of all of my many ancestors; and the list could go on and on. The point is that there are countless factors needed for me to be in the exact position in which the racket and ball come together.

So at the deepest level, the racket itself didn't cause the ball to go over the net. Both the racket and ball are products and parts of the universe (consciousness). They are emergences of the universe itself.

As stated by Maharaj: "Like everything mental, the so-called law of causation contradicts itself. No thing in existence has a particular cause—the entire universe contributes to the existence of even the smallest thing; nothing could be as it is without the universe being what it is. When the source and ground of everything is the only cause of everything, to speak of causality as a universal law is wrong."[58]

Dr. David Hawkins put it another way: "The belief in linear causality is a basic axiom of the whole structure of the ego/mind dualistic belief system. To see through that illusion is the most important and greatest leap available for getting closer to comprehending Reality."[59]

Causality, as we know it, also runs into problems with our new understanding of "time." The common-sense notion of causality implies that some event must come *before* another event to cause it. Put another way, it implies that time goes from past to present to future. But findings in quantum physics are beginning to question that basic assumption. The findings suggest that the future may impact the past. Furthermore, we've seen examples of precognition that put into question the direction of the arrow of time.

Time as we conceive it might be an illusion (more on this to come). And if that is the case, then perhaps the basic concept of "X from the past causes Y in the future" doesn't apply at the most fundamental level.

If consciousness is, in fact, fundamental, maybe it simply *exists*—beyond space and time—without anything acting as the cause for it to be. It just *is*. The notion that something needs to come from something else in order to exist might just be an assumption of the inherently limited human mind. Causality appears to be true in most localized events we see and experience, but what we see and experience is extremely limited relative to the whole.

Dr. Eben Alexander puts this notion into words when he recalls his life-changing NDE: "I had no real center of consciousness. I didn't know who or what I was, or even *if* I was. I was simply…*there*, a singular awareness in the midst of a soupy, dark, muddy nothingness that had no beginning and, seemingly, no end"[60] [emphasis in original].

What existed before consciousness?

If time is, in fact, illusory, then perhaps there was no "before" consciousness. It exists *beyond* space and time. Hence, we see Dr. Alexander's representative NDE description of feeling like his consciousness was infinite and without beginning or end. Nobel Prize-winning physicist Erwin Schrödinger even stated, "There is really no before and after for the mind. There is only now that includes memories and expectations."[61] He also commented, "We may, or so I believe, assert that physical theory in its present stage strongly suggests the indestructibility of Mind by Time."[62]

Maybe consciousness just *is*, even if that's difficult for our human minds to fully grasp. Just because something is difficult to grasp doesn't make it unreal. For example, the concept of "infinity" is widely accepted. But can we fully grasp what infinity is? No.

If consciousness simply *is*, and is all that is, then each of us would be part of that consciousness. Therefore, we would all be connected. Are we, in fact, all connected?

The idea that we are all connected seems plausible based on the evidence we've seen. For example, recall from chapter 3 the concept of quantum entanglement. In quantum entanglement, the states of two physically *distant* particles mirror each other *instantaneously*. This phenomenon points toward the existence of invisible connections that we can't see with our eyes.

NDErs such as Anita Moorjani sometimes describe a feeling of oneness. She recalls feeling that she "encompassed—no *became*—everything and everyone"[63] [emphasis in original]. This idea also appears during some life reviews in NDEs. Recall a passage from chapter 9 in which an NDEr describes a life review: "Not only did I perceive everything from my own viewpoint, but I also knew the thoughts of everyone involved in the event, as if I had their thoughts within me. This meant that I perceived not only what I had done or thought, but even in what way it had influenced others, as if I saw things with all-seeing eyes."[64] This notion makes sense if each of us is fundamentally interconnected as part of the same underlying consciousness.

Furthermore, psychic phenomena suggest a hidden interconnectedness among minds. Dr. Radin goes as far as to posit an "entangled minds" theory, arguing that these phenomena are "an unavoidable consequence of living in an interconnected, entangled physical reality."[65]

Physicist Nassim Haramein provides a similar perspective in his 2016 essay *The Physics of Oneness*: "We live in a highly entangled, interconnected universe where a fundamental information field is shared across all scales to generate organized matter and eventually self-organizing systems, leading to organisms reflecting back on themselves asking fundamental questions about existence. This process drives evolutionary mechanisms in which the environment influences the individual and the individual influences the environment, in a nonlocal interconnected wholeness—a universe that is ultimately ONE."[66]

And recall that Nobel Prize-winning physicist Erwin Schrödinger even explicitly stated: "In truth, there is only one mind."[67]

The idea that we are all one sounds nice, but I don't feel it. I look around and see objects and people who are separate from me. I'm me. They are them. How could it be that we are all one?

Let's do an exercise (inspired by and adapted from Rupert Spira's teachings) that will require close inspection of your own experience. Go as slowly through this segment as needed.

> Imagine that you see a tree in the far distance. We are conditioned to think: "I am here, and there is a separate tree over there." But is that accurate?
>
> What do you truly know of the tree? The only thing you truly know of the tree is your experience of it. And your experience of it is that you see it. It is a visual experience.
>
> Now I ask: "Where does your experience of seeing it occur? Does seeing the tree occur 'here' or does it occur 'over there'?"
>
> In fact, you realize that your seeing of the tree occurs right here, within your field of consciousness. In other words, what you have been labeling "tree over there" is really just a visual perception that occurs "here" rather than "over there." The tree is nothing more than a modulation of your field of consciousness. So the tree is made of consciousness. The tree is just in your mind.
>
> You might then wonder: "Well, I can walk to the tree and touch it, and that touching will occur over there and not here." When you do this, your hand touches the bark of the tree. You feel a tingling sensation that you label "hand touching bark." But if you close your eyes and lose all association with the concepts

of "hand" and "bark," what do you feel? Imagine you were a newborn infant who didn't know what any objects were. How would the infant describe the sensation of what you call "hand touching bark"? The answer is: All that you experience is a tingly sensation. As you close your eyes and imagine this, *where* does that tingly sensation occur? The sensation occurs right in your field of consciousness. The sensation of touching the tree is nothing more than a modulation of your consciousness. The tree—as you "see" and "feel" it—is nothing but consciousness. The tree exists in your mind.

You might then wonder again: "What if I hear one of the tree's branches break—the sound occurs at the tree, which is "over there." Run the same exercise as above. Where does your experience of hearing the branch break occur? Does it occur at the site of the tree? Close your eyes and imagine it; think deeply about where your experience of hearing the branch break occurs. You will find that just like the seeing of the tree and the feeling of the tree, your hearing of the branch break occurs in your field of consciousness. The sound is a modulation of your consciousness. It is nothing but consciousness. The tree—as you "see" and "feel" and "hear" it—is nothing but consciousness. The tree only exists in your mind. There is no "tree" independent and outside of your mind. And since the tree exists in your mind as a modulation of your consciousness, it really is just a modulation of *you*. You aren't separate from the tree, even though you've been conditioned to interpret your perceptions that way.

Similarly, consider thoughts and emotions. Where do they occur? Close your eyes, and watch your thoughts and emotions flow. You will find that they, too, occur in your field of consciousness. They are nothing but modulations of your consciousness. They are made of consciousness.

Consider even your body. Forget for a moment any labels, and act as if you were an infant who knew nothing of worldly definitions and instead only knew pure experiencing. Your body is a set of sensations that you feel and a set of perceptions you see and label as "body parts." Where do those sensations and perceptions of your body occur? They occur here in your field of consciousness. They are nothing but modulations of consciousness. They are made of consciousness. Your body is made of consciousness. Your body exists in your mind.

This exercise could be done for *any experience of anything*. Our life is nothing but a set of experiences; therefore, everything in our life can be thought of in this way. Any thought, emotion, sensation (such as feeling, tasting, smelling), or perception (seeing, hearing) occurs in your field of consciousness. They are all modulations of your consciousness. They are made of consciousness. Everything is consciousness and therefore everything is one.

As further summarized by physicist Peter Russell: "All that I perceive—everything I see, hear, taste, touch, and smell—has been reconstructed from sensory data. I think I am perceiving the world around me, but all that I am directly aware of are the colors, shapes, sounds, and smells that *appear in the mind*"[68] [emphasis added].

(If you are finding this concept difficult to wrap your head around, you are not alone. It requires an unwinding of a lifetime of conditioning. These exercises take time and practice to truly feel and experience. In fact, Rupert Spira has created a 30-hour meditation set called *Transparent Body, Luminous World: The Tantric Yoga of Sensation and Perception*, which is designed to allow the body to feel these concepts.)

I'm still confused. How is everything "in the mind" when we know we are in a reality that at its core comprises space, time, and matter?

As we saw in chapter 3, space, time, and matter are less fixed than we thought. Einstein's relativity theory shows that space and time shrink or dilate depending on relative speed and gravitational force. We've seen in this book that time may not go from past to present to future. Furthermore, we've seen that "matter" is almost entirely empty space and that it exists as a wave unless an observer observes it. As stated by Dr. Laszlo: "When physicists descend to the ultrasmall dimension, they do not find anything that could be called matter."[69]

So we know that space, time, and matter work in ways that go against our ordinary, everyday experiences. If we want to claim that our world is made of "space," "time," and "matter," we should take a step back and define what they are.

In chapter 3, we established that we don't know what matter is fundamentally made of. We don't know what it is.

What about time? As Dr. Bernardo Kastrup suggests, we could define time as "the interval between two events." But "interval" is just another form of "time." So we're using time to define time. That doesn't help! Can we find another way of defining it so that we don't use the word we are trying to define in the definition? No, we can't. When we try to define time, we end up in a circular loop that gets us nowhere.[70] We can't meaningfully define it.

Dr. Kastrup provides a similar example with regard to "space." If we say that space is merely "the distance between two objects," we run into the same problem we had with "time." Using "distance" to define "space" is circular. If you define "distance," you realize its definition involves "space." And we don't want to define "space" using the term "space." So, like time, we can't define space with language either.[71]

As stated by Dr. Kastrup:

> We all take space and time for granted until we try to tell ourselves what they are. We then discover that, despite the fact that we seem to inhabit them, they can't be defined without reference to themselves. They arise magically from self-reference.... Space and time are like ghosts that vanish into thin air every time we try to grab them. Their "form" is "emptiness" referring to itself in a kind of cognitive short-circuit. Indeed, *if you can't tell yourself what something is, then it's most-likely an illusion resulting from a cognitive short-circuit*; it isn't really out there. More specifically, I suggest that space and time are language ghosts. They only seem to exist as independent entities because we conceptualize them in words[72] [emphasis in original].

I'm still confused—in my experience, past events have occurred, and I have been to many different locations. So I have been to different points in time and space. I experience time and space every moment of every day. How can you claim that they are illusions?

You claim that past events occurred. For example, you woke up this morning. You had lunch yesterday. You graduated from high school. You had a clown at your fifth birthday party. You were born. All of these events occurred in the past.

Can you *prove* that those experiences occurred in the "past"? Have you ever actually been to, or experienced, the "past"? No, you can't prove that events occurred in the past, and you haven't ever been to the past. The past is an inference. All you truly know of the past are your memories or stories about the past. But those memories and stories occur *now*. They always occur *in the present moment*. In the words of Dr. Kastrup: "The same goes for historical artifacts in...museums: we perceive them in the present. That they come from the past is a story—an explanation—whether true or false. The past is always a myth."[73]

Similarly, have you ever been to the future? All we ever know of the future is our thoughts about the future. And those thoughts occur *now*. They always occur *in the present moment*.

Has it ever not been now? Nope. It has always been, and always will be, the present moment.

What is the present moment exactly? We can't even define it. If we try to identify when the present moment happens, by saying, "It's the present moment," we are a split second too late. It is occurring, but we can't quite capture it.

Dr. Kastrup therefore concludes: "Past and future are myths: stories of the mind.... You cannot escape the present; ever; not even theoretically."[74] Further, he states "that we think of life as a series of substantial happenings hanging from a historical timeline is a fantastic cognitive hallucination."[75]

With this understanding, it becomes clear why many philosophers talk about the importance of focusing on "the now," as in Eckhart Tolle's #1 *New York Times* bestseller entitled *The Power of Now*.

Let's do the same exercise as before, but only now with space.

> For example, you woke up this morning in your bed. You had lunch yesterday at a restaurant. You graduated from high school in your hometown. You had a clown at your fifth birthday party at your childhood home. You were born in the

hospital. Those all seem like events that took place at different locations. In other words, in all cases you were in different points in space.

But were you, really? If I asked you right now, "Where are you?" an accurate answer would be "here." When you woke up in your bed this morning, at that moment, you were "here." When you had lunch yesterday, at that moment, you were "here." You could continue this exercise for any event in your life. In all cases you would describe yourself as being located "here." Similarly, in all cases you would describe the time (at the time of the event) as having been "now."

You might contend, "But I can see a 'there' over there, and that is not 'here.'" As discussed earlier, your perception of "there" occurs "here" in your field of consciousness.

As Spira phrases it: "The experience of 'there' takes place *here*, just as the experience of the past or future takes place *now*. It is not possible to leave 'here' and visit 'there.' 'There' is always a concept, never an experience. Space is the distance between the point 'here' and the point 'there,' or between two points 'there.' However, only *here* is experienced"[76] [emphasis in original].

But where is "here"? Can we pinpoint it? No, we can't. It's like a place with no definable location. Similar to "now," we can't truly define or isolate "here."

And yet you are always "here." You've never not been "here." Similarly, it's always been "now." It's never not been "now." Table 1 below illustrates that it has only ever been here and now. From the point of view of the perceiver at the moment of the event, it is always here and now.[77]

Event (illustrative only)	**Time at which event occurred (at the moment of the event)**	**Location of the event (at the moment of the event)**
You woke up this morning in your bed.	Now	Here
You had lunch yester-day at a restaurant.	Now	Here

Event (illustrative only)	Time at which event occurred (at the moment of the event)	Location of the event (at the moment of the event)
You graduated from high school in your hometown.	Now	Here
You had a clown at your fifth birthday party at your child-hood home.	Now	Here
You were born in the hospital.	Now	Here

So what are space and time, really? We established that they can't be defined in any truly useful way. And the only thing we can definitively say about our experience of these undefined terms is that they are "here" and "now," respectively. And on top of that, we can't even pinpoint "here" and "now"; they are totally amorphous and borderless.

Princeton PEAR lab's Dr. Robert Jahn and Brenda Dunne recognized the inherent issues with space and time through their nearly 30 years of nonlocal consciousness research. They commented that their studies raise the question "of whether space and time are truly intrinsic properties of the physical world, or whether, as many prominent thinkers have maintained, they are *subjective* coordinates that consciousness imposes in order to organize its experiences"[78] [emphasis in original].

In other words, the physical world of "matter" appearing in "space" and "time" isn't as we think it is. All are just constructs of the mind. As Spira sees it, time is a way in which the human mind interprets thought. Space is the way in which the mind interprets perception.[79] He flips our normal conception of space and time on its head: "We are not moving through time and space. Time and space are, as it were, moving through us."[80] Further, he comments, "I never go anywhere. I am always in the same place of 'I am,' the placeless place called here, the timeless time called now."[81]

Space and time (and matter) are just in the mind; they are in consciousness. For this reason, we can more fully understand why Johns Hopkins physicist Richard Conn Henry commented: "The Universe is mental."[82]

And we can see why mystics such as Ramana Maharshi is quoted as saying: "The world one sees does not even exist."[83]

If the world we see—consisting of space, time, and matter—does not exist outside of the mind, where do I fit into this? Who am "I"? I exist, right?

An obvious part of your existence is that you are here, right now, having an experience. Whether space, time, matter, or anything else is "real" is up for debate, but it is undeniable that you are reading these words, having an experience as the "I" who is reading them. "I" indisputably exists as the experiencer right here and right now. It is your ultimate identity.

Our states of being can vary. You might say: "I am happy"; "I am healthy"; "I am in California"; "I am a wife"; "I am an artist"; "I am a high school graduate."

Whatever the description, there is a constant. That constant is "I am."

Who is "I"? Let's investigate by walking through an exercise inspired by the teachings of Rupert Spira.

Can we define "I" as our thoughts? Right now, you are thinking about this book. Earlier you were thinking about your family. Before that you thought about your job. Your thoughts change. But the "I" that was having the thoughts did not change. "I am thinking about this book"; "I am thinking about my family"; "I am thinking about my job." "I" has always been there during your sea of changing thoughts. Therefore, we cannot equate "I" with your thoughts. "I" and thoughts are not one and the same. "I" experiences the thoughts, but is not made of the thoughts.

Is "I" my body? My body today is not the same as it was five years ago. It is not the same as it was 20 years ago. It is not the same as it was when I was an infant. But in all cases, "I" was present. "I" hasn't changed, but my body has. Dr. Kastrup comments: "If I lost a limb or had a heart transplant tomorrow I would still have the same sense of identity."[84] The "I" that existed before these bodily changes is the same as it is afterward. Therefore "I" cannot be equated with my body. "I" experiences the body, but "I" is not the body itself.

Is "I" my genes? We know from biology that genes mutate. So I don't have the exact genetic makeup I had earlier in my life. Yet the same "I" persists even if my genes are slightly different. Thought of another way by Dr. Kastrup: "Am I my genetic code? No, for I could have an identical twin with the same genetic code and I wouldn't be him."[85] So there is a distinction between "I" and my genes. "I" is not the same as my genes.

This exercise could go on and on. In any instance of "I am ____," we will see that there is a distinction between "I" and whatever the state "____" is. Try it—insert anything into "____," and see if you can find something that is identical to "I." It can't be done.

As stated by Dr. Hawkins: "Just as the eye is unaffected by what is observed or the ear by what is heard, there is the ongoing process of witnessing, which is unaffected by that which is witnessed."[86] The "I" that is witnessing one's life persists while states of its experience may change.

So we have just established that "I" is untouched by any transitory states that come after "I am." "I" is without qualification or limitation. "I" is *un*-limited.

We then might ask, "Who is it that knows that 'I am'?" The answer: "I." Is the "I" that knows that "I am" the same "I" as the one in "I am"? Yes. So "I" is aware of itself. "I" knows itself. "I" is both the perceiver and perceived, the knower and the known, the subject and the object. "I" is therefore self-aware.

We might then wonder what "I's" form is. My body has a measurable form. My table has finite boundaries. I can measure the exact size and boundaries of my cell phone. But does "I" have any borders? Close your eyes and think about it. Can you pin down the "I" that is having experiences? Can you put a boundary or border on it? Can you confine it to a definitive space? The answer you will come to is that you can't put anything finite around "I." It's not a finite thing, whatever it is. "I" is therefore *in*-finite.

Finally, we might wonder when "I" is present. Right now, "I" is present reading these words. Ten years ago, "I" was present. When I was born, "I" was present. As far as we can tell, has "I" ever *not* been present? No. "I" has *always* been present in our life. At the very least, from birth to death, "I" is present. And we haven't experienced our own birth or our own death; in order to experience such hypothetical events, we would need to do the impossible of being present *before* birth and *after* death. As far as our experience tells us, "I" is always present. "I" has never not been present. Another way of phrasing its ever-presence is to say that "I" is eternal.

So, our own introspection tells us that "I" has the following characteristics: *It is unlimited, self-aware, infinite, and eternal* (see the following table for a summary).

Characteristics of "I"	Rationale
Unlimited	"I" can experience many different states, but those states change while the same "I" remains. If I say, "I am ____," no matter what "____" is, "I" is present. There is no limit that can be placed on "I."
Self-aware	"I" is the entity that is aware that "I am"; it is aware of itself.
Infinite	"I" does not have a finite boundary like my body or my table. "I" is borderless and uncontained.
Eternal	"I" is always present in my life, no matter what state I am in. "I" is ever-present (i.e., eternal).

Religions—across cultures—have a name for an unlimited, self-aware, infinite, and eternal being. The name they give for that being is: "God."[87]

Now we can truly understand why the 13th-century Sufi mystic, Rumi, said: "I searched for God and found only myself. I searched for myself and found only God."

And we can understand why those who experience transcendental states of consciousness (e.g., NDErs and some psychedelics users) claim to experience divinity: They are getting in touch with the "I" that they truly are. As Anita Moorjani recalls from her NDE, "I was overwhelmed by the realization that God isn't a being, but a state of being…and I was now in that state of being."[88]

Is "I" the same for each of us?

If we performed the above exercise for every living being, we would find the exact same qualities. "I" that is experiencing the life of Mark has exactly the same characteristics that any other would give to his or her "I." Isolating "I" leads to the same conclusion about its characteristics: It is unlimited, self-aware, infinite, and eternal. All of us, as "I," have those exact qualities.

The implication is that the "I" that is me is the same as the "I" that is you. Our bodies, thoughts, feelings, sensations, perceptions, and experiences might be different, but the "I" that experiences them is the same.

Notions of "God" and "we are all one" and "equality" now take on profound new meanings. Our focus shifts inward rather than outward. We are guided toward what we are rather than what we are seeking outside ourselves.

Dr. Kastrup comments: "The conclusion…is that our inner sense of 'I' is fundamentally independent of any story we could dress it up with. As such, it is entirely undifferentiated and *identical* in every person"[89] [emphasis added].

And Spira states: "Each of us is, as such, the *same person*, apparently diversified and separated through the kaleidoscope of thought and perception.…We are literally each other. Each of us is the outer face, or the objectification, of the only mind there is, eternal, infinite consciousness. We are all mirrors of the same consciousness"[90] [emphasis in original].

Finally, Nobel Prize-winning physicist Erwin Schrödinger echoed these sentiments: "To divide or multiply consciousness is something meaningless. In all the world, there is no kind of framework within which we can find consciousness in the plural; this is simply something we construct because of the spatio-temporal plurality of individuals, but it is a false construction."[91] Hence, as mentioned earlier, he stated: "In truth, there is only one mind."

Would this notion change our conceptions of love and beauty?

It certainly gives a way of recontextualizing them. When I was a psychology student at Princeton, a common approach we took to understanding human behavior was to uncover the evolutionary basis for it. We would ask, "In what sense did this behavior enable an organism to survive so that it could reproduce?" In many instances, we were able to come up with good reasons. For example, the behavior of lust makes a lot of sense. Sexual reproduction is required for the continued survival of a species, so the desire to have sex is critical and would be selected for, and passed on to, offspring.

But a question that continually came up in my psychology classes was: "What is the evolutionary explanation for love?" It was a mystery. We could come up with reasons why love would be helpful. Caring for others might help the chances of survival and enable the passing along of genes. But is love *required* for that? Debatable. Certainly not a slam dunk. And yet love is a fundamental part of the human experience, but science doesn't fully understand it.

Similarly, there aren't strong evolutionary explanations for natural beauty. Why would it be evolutionarily beneficial to find a sunset beautiful? Or a painting? Finding things beautiful wouldn't confer a selective advantage for survival, would it? Yet the notion of beauty is pervasive in the human experience.

Perhaps there is another explanation that takes evolution out of the equation. What if love and beauty have something to do with our innate quality as consciousness—as "I"—a quality that transcends biology. As Spira puts it: "At the deepest level all minds are connected because they are all precipitated within the same field of infinite consciousness, and the varying degrees of connectedness that we feel with one another or with animals, objects and nature are the degrees to which our minds are transparent to this shared medium. Love is the word we use when we *feel* this shared medium with other people and animals. The same experience is referred to as beauty in relation to objects"[92] [emphasis in original].

And if we are just part of a singular consciousness—the singular, shared medium of existence—perhaps love and beauty are our intrinsic characteristics. So they aren't products of evolution or biology; they are basic to what we fundamentally are!

With this lens, we can perhaps explain feelings described in mystical experiences such as those induced by psychedelics and NDEs. Experiencers typically describe feelings of unconditional love, and they appreciate beauty in new ways. These experiences occur when the brain has *reduced* functioning—meaning that the experiencer is *more* tapped into the broader consciousness and its naturally loving and beautiful state of being. In these experiences, the brain's filter is partially unlocked.

So perhaps we experience our natural state of love and beauty *more* when our brain does *less*.

Are these experiences similar to what is described as "enlightenment" or "awakening"?

Certain individuals throughout history have reported states of "enlightenment" or "awakening" in which they achieve the above-mentioned loving and blissful state through meditation or other (sometimes spontaneous) means. From a materialist standpoint, claims of these states must be biological side effects of the brain that we don't yet fully understand. They occur because of chemicals in the brain and nothing else.

If we view consciousness as being fundamental, then maybe enlightenment and awakening are something more. Maybe they have something to do with aligning with our "true" self; the "I" within us that we have for so long overlooked. Maybe they are additional experiences of "unlocking the filter" so that we experience the broader reality that our brain normally hides from our experience.

Consider the case of Sir David Hawkins MD, PhD, a well-respected psychiatrist who authored *Orthomolecular Psychiatry* (1973) alongside Nobel Prize winner Linus Pauling. During his life, he reached states resembling what many describe as enlightenment. He described those states: "It was necessary to stop the habitual practice of meditating for an hour in the morning and then before dinner because it would intensify the bliss to such an extent that it was not possible to function."[93]

He also reported apparent miracles:

> The miraculous happened, beyond ordinary comprehension. Many chronic maladies from which the body suffered for years had disappeared; eyesight spontaneously normalized, and there was no longer a need for the lifetime bifocals. Occasionally, an exquisitely blissful energy, an Infinite Love, would suddenly begin to radiate from the heart toward the scene of some calamity. Once, while driving on a highway, this exquisite energy began to beam out of the chest. As the car rounded a bend, there was an auto accident; the wheels of the overturned car were still spinning. The energy passed with great intensity into the occupants of the car and then stopped of its own accord. Another time while I was walking on the streets of a strange city, the energy started to flow down the block ahead and arrived at the scene of an incipient gang fight.

> The combatants fell back and began to laugh, and again, the energy stopped.[94]

Dr. Hawkins found that "people felt an extraordinary peace in the aura"[95] of his presence.

He also stated of the experience: "The novelty of sequential experience disappears as do expectation, regret, or the desire to anticipate or control. Existence as Existence is total and complete. All one's needs are already fulfilled. There is nothing to gain or lose and everything is of equal value. It would be like all movies being equally enjoyable because the pleasure stems from going to the movies, and the movie that is playing is irrelevant."[96]

In another case, philosopher Eckhart Tolle had an unplanned "awakening" in which he reached similar states. After suffering from anxiety and depression, he describes the transformation he underwent: "The nightmare became unbearable and that triggered the separation of consciousness from its identification with form. I woke up and suddenly realized myself as the I Am and that was deeply peaceful." Since then, his peace and bliss have remained: "Basically, the peace is continuously there. There is a variation of the intensity. At first it was an intense experience for a long period of time—weeks, months, years. It was a kind of bliss, but it was only bliss in contrast to what had been before. Now that kind of peace is normal. Once bliss becomes normal it's no longer bliss, it's just peace."[97]

Dr. Edith Ubuntu Chan, holistic medicine doctor and Harvard graduate, had a similar spontaneous awakening while in meditation. As she describes it: "One moment, I was sitting peacefully following a guided…meditation. Then the next moment…I experienced myself bursting…Bursting into trillions of pieces of Love and Light. I experienced myself the size of the entire cosmos. No more physical body. No time. No space. The feeling was so intensely blissful, so beautiful, so filled with love…I had no reference point for anything like this in my earthly life. All I knew was that—I was Home. All I knew was that—this is our natural state."[98]

Transpersonal psychologist Dr. Bonnie Greenwell had a similar experience:

> An intense spasm suddenly rolled up my spine and through my head. My body jerked upright as rivulets of bliss moved through my nervous system. It was like being electrified with

> joy. The energies pushed upward over and over, each roll more intense than the one before, my mind becoming progressively less focused. After a few minutes I was barely sitting, listening to a college lecture on developmental psychology. I did not know it then but the direction of my life, my interests and my work was being blasted into new territory and I would never see the world again through the same lens.
>
> When the class was over I stumbled down the hall and into a small meditation room....I sat on a cushion leaning against a wall and fell into a vast and open sense of spaciousness, my mind empty and quiet, my body completely lost in a floating pleasure. Eventually I came back into my senses, but for weeks following I was in a state of awareness that was untouched by the normal challenges of my family, studies and work, even though they continued as usual. Walking down the street felt like floating. Waking in the morning felt like entering a new adventure. Sitting to meditate felt like falling into a world of light and bliss. At night my body would awake and shake itself, move into positions that stretched the spine, plunge my nervous system into a vibration of happiness, and occasionally produce an other-worldly dream.[99]

Dr. Greenwell has since focused her life's work on helping people who have had awakening experiences such as what she experienced—what she and others call "kundalini awakenings." These awakenings involve a primal kundalini energy theorized to exist at the base of the spine. It has been discussed in Eastern traditions, but it is not generally accepted by Western medicine.

As Dr. Greenwell summarizes it: "The activation of kundalini energy has been embraced by some schools of yoga in the West, and often marketed as a way to enhance health and longevity. But its true function is to shift a person's consciousness, to awaken his or her true nature, and it does this through a process that brings a deconstruction of the old self and collapses the familiar identifications. It may happen that health and longevity is improved."[100]

Given the intensity of Dr. Greenwell's and others' experiences, the topic seems worthy of further investigation by more scientists. We all want to be happy. We all want to feel good. Enlightened or awakened states appear to induce immensely pleasurable changes that are associated with a shift in one's consciousness. Consistent in these experiences is an inward shift, an identification with one's true self rather than the external world, a focus on the formless but ever-present "I."

What are the implications for happiness?

Our culture emphasizes the need to look outside ourselves to find happiness. We seek the next relationship, the next financial success, the next meal, the next toy, the next high, the next vacation, and so on. Once we achieve our goals, there is often temporary happiness and a short-term rush, but the positive feelings dissipate over time. We return near our baseline happiness level and then look to chase after the next object. For example, if external objects really did bring happiness, then we might expect rich people to be consistently happier than poor people (assuming that we are speaking of poor people whose basic needs are met). The research does not show that, however. Rich people and poor people are just as happy (although rich people experience less sadness).[101] Intuitively this makes sense. We often hear of rich and famous people who ostensibly have it all (by material means), but they are depressed, addicted to drugs, and are still seeking true happiness.

Once we get what we want, there is a tendency to adjust to it. It starts to lose its allure. The buzz wears off, so we search for the next "high." Psychology has a term for this phenomenon: it is known as the "hedonic treadmill." It's as if we are on a treadmill running for the next object outside ourselves, but not actually going anywhere. We feel like we will get somewhere because we are running, but the treadmill keeps us in place.

Spira summarizes the phenomenon well: "The person wrongly attributes the peace, happiness, and freedom briefly experienced to the acquisition of the object, activity, substance, state of mind, or relationship and, as a result, when the underlying suffering resurfaces again in between the normal activities of the outward-facing or object-seeking mind, he simply returns to the same objective experience, hoping thereby to experience the same relief, in an ever-deepening cycle of longing, addiction and despair, each time requiring a slightly stronger dose of the object to achieve the desired result."[102]

Or as stated by Eckhart Tolle: "Even if you achieve your outer purpose, it will never satisfy you if you haven't found your inner purpose, which is awakening, being present, being in alignment with life."[103]

Spira often provides an analogy that is instructive here. Imagine a colorless, self-aware television screen that extends out into infinity. Characters and objects appear as movies on the screen. They are colorings of the screen, but they are *made* of nothing more than the screen itself. The characters get lost in the story of the movie. All the while, they've forgotten they are fundamentally nothing more than the screen itself! Are we, the characters in the movie, forgetting that we are nothing more than the ever-present "I" of the "screen"? The character we are is a collection of thoughts, perceptions, sensations, and feelings; however, as we established, "I" is constant while all forms of experience change within it. The failure to look inward causes us to seek illusory objects in the movie at the expense of ignoring that which perceives: "I."

In light of these ideas, we should consider a statement made by the 14th Dalai Lama:

> In this century, human knowledge is extremely expanded and developed. But this is mainly knowledge of the external world. In the field of what we may call "inner science" there are many things, I think, that you do not know. You spend a large amount of the best human brain power looking outside—too much—and it seems you do not spend adequate effort looking within. Perhaps now that the Western sciences have reached down into the atom and out into the cosmos finally to realize the extreme vulnerability of all life and value, it is becoming credible, even obvious, that the inner science is of supreme importance. Certainly physics designed the bombs, biology: the germ warfare, chemistry: the nerve gas, and so on, but it will be the unhealthy emotions of individuals that will trigger these horrors. These emotions can only be controlled, reshaped and re-channeled by technologies developed from successful inner science.[104]

Maybe we should heed the Dalai Lama's advice and focus our attention inward. Modern society doesn't teach us to do that. We are accustomed to looking externally. We are conditioned to look for the next best thing outside of ourselves.

Are we, as a society, just like Dorothy in the 1939 movie *The Wizard of Oz*, who tirelessly searched the land of Oz to find her way home to Kansas only to discover that she had the power to go home the whole time? At the end of the movie when she realizes she had the ability within her all along, Dorothy says: "If I ever go looking for my heart's desire again, I won't look any further than my own backyard, because if it isn't there, I never really lost it to begin with."

What are the implications for world peace?

If we are indeed what the evidence in this book suggests we are—an entangled, interconnected stream of infinite consciousness with no actual separation, with no beginning and no end—how might we act toward ourselves and others? How might we run businesses? How might we lead nations?

But the dominant paradigm in today's society isn't one of interconnectedness. As Spira often reminds us in his teachings, materialism is the dominant paradigm, and it implies that we are finite and limited beings who are born and destined to die. It also teaches that we are fundamentally separate. My consciousness is separate from your consciousness. Maybe you've never explicitly thought about it because it's such a core assumption in our society. But the implication of materialism—which suggests that the brain produces consciousness—is undoubtedly that we are finite and separate.

I view the belief that we are finite and separate to be the disease underlying virtually every problem in human society today. Anxiety, depression, interpersonal problems, racial and social prejudices, gender inequality, geopolitical unrest, violence, war, greed, or nearly any problem you can think of—at their core are symptoms, not the disease. The world's problems are caused at the most fundamental level by the pervasive underlying assumption that we are all finite, limited, and separate. And that stems from the materialist belief that consciousness comes from the brain.

Dr. Kastrup summarizes the inescapable implications of materialism: "By linking consciousness and personal identity to limited and temporary arrangements of matter, materialism inculcates the following subjective values in our culture: life is short and you've only got one to live; the only source of meaning lies in matter—after all nothing else exists—so the game is to accumulate as many material things as possible; we should consume as fast as possible, even at the expense of others or the planet, for we have nothing to lose since we're going to die soon anyway."[105]

With such a toxic underlying belief system, is it any wonder that the world seems to be falling apart in front of our eyes? For this reason, Spira states:

"If humanity does not still exist in five hundred years' time, it will most likely be because materialism prevailed. Humanity cannot survive the materialist paradigm."[106]

Think for a moment. What if, as this book suggests, the unprovable religion of materialism is wrong? What if the brain does not produce consciousness? What if consciousness is the basis of reality, and my consciousness, "I," is the same as yours, which is the same as everybody's? What if we are the same consciousness, simply looking at our "self" through different eyes?

If materialism is wrong, and we are not finite and not separate, would we treat ourselves or others poorly? Would you want to harm another if you knew that person or animal or insect or anything was just another version of you? Would it be rational to harm yourself?

Altruism—a concept that has for many years troubled evolutionary biologists and psychologists—makes so much sense under this framework. Of course it feels good to help another—by helping another, we are helping our "self" as part of the same underlying consciousness. In this sense, altruism is the highest form of selfishness.

These ideas are nothing short of world-changing.

As Spira eloquently suggests: "Consciousness is the fundamental, underlying reality...and...the overlooking, forgetting or ignoring of this reality is the root cause of both the existential unhappiness that pervades and motivates most people's lives and the wider conflicts that exist between communities and nations. Conversely,...the recognition of the fundamental reality of consciousness is the prerequisite and a necessary and sufficient condition for an individual's quest for lasting happiness and, at the same time, the foundation of world peace."[107]

Dr. Larry Dossey makes another well-stated plea in his 2013 book, *One Mind*: "One Mind...is a collective, unitary domain of intelligence, of which all individual minds are a part....I believe the One Mind is a potential way out of the division, bitterness, selfishness, greed, and destruction that threaten to engulf our world—from which, beyond a certain point, there may be no escape. Identifying with the highest expressions of human consciousness can clear our vision, prevent the hardening of our moral and ethical arteries, and inspire us to action. These are not ordinary times. Urgency is afoot."[108]

These statements are not hyperbole. Given the state of affairs in the world today, it seems that there is no more important study than the study of consciousness—for the sake of science, technology, medicine, business, education, ethics, and politics—and for the sake of humanity.

Let the next scientific revolution begin.

Acknowledgments

> *When I look at any achievement I find that it is there because of certain opportunities I had, as well as because of my personal effort. I cannot claim to have created or commanded the opportunities; they were given to me. I happened to find myself in the right circumstances, and so I could grow and learn what I needed to learn. I met with the right person; happened to read the right book; I enjoyed the right company; someone came forward with the right guidance at the right time. There are so many factors behind an achievement. I cannot really say I created any of them. When I look at the facts, I must see that any achievement that I claim as mine is not due to my will or skill alone but it is due to certain things and opportunities that were provided to me.*
>
> —Swami Dayananda Saraswati, *Value of Values* (2007)

While my name is on the cover of this book as the author, it would not have been possible without the support and guidance of so many people. I like to keep in mind the advice of the late Dr. David Hawkins, who suggested that instead of being proud we should be grateful.

My literary agent, Bill Gladstone, deserves all the credit in the world. He quickly recognized the potential impact of this book and took a chance on me as a first-time author. I could not have asked for a more supportive and philosophically aligned agent. And I have had the even greater pleasure of working with Bill and his wonderful wife, Gayle, as my publishers at Waterside Press. Thank you for believing in me.

Thank you to Lisa Barnett and Alexis Sclamberg for independently suggesting that I approach Bill Gladstone with my manuscript.

It has been a privilege to work with Waterside's Editorial Director and Managing Editor, Kenneth Kales. Kenneth's world-class editorial skills played a crucial role in refining my manuscript. Additionally, it was an honor to work with Jill Kramer, a highly talented and experienced editor and proofreader who helped put the finishing touches on the book. Many thanks to Joel Chamberlain for his assistance with the image design, bibliography, and typesetting. Thank you to Ken Fraser for such a fantastic cover design. Thank you to Tray McCurdy for taking such high-quality headshots. I thank my terrific publicists Barbara Teszler, Jackie Lapin, and Dalyn Miller. And thank you to Jennifer Uram for her support with my book-related contracts.

Dean Radin and Rick Hanson were kind enough to review the first draft of my manuscript and encouraged me to bring the book to a mainstream audience. I am grateful to the many individuals who provided feedback on various versions of my manuscript, all of which made the book better: Dean Radin, Ed Kelly, Elissa Epel, Julia Mossbridge, Jude Currivan, Roger Nelson, Jim Tucker, Tyler Heishman, Robert Hellauer, Casey McCourt, Tray McCurdy, Eric White, Kylie Richardson, Danny Oppenheimer, and Natalie Carlstead.

Many thanks to those who have offered such thoughtful endorsements and support (as of July 2018): Goldie Hawn, Ervin Laszlo, Dean Radin, Larry Dossey, Eben Alexander, Jack Canfield, Julia Mossbridge, Tiffany Pham, Brenda Dunne, Jude Currivan, Ed Kelly, Loren Carpenter, Edith Ubuntu Chan, Giancarlo Marcaccini, Barry Baker, Gregory Miller, Roger Nelson, Elissa Epel, Ann Shippy, Rupert Sheldrake, Guru Singh, and Stephen Sinatra.

The personal journey that led me to write this book has been supported by an amazing group of individuals who only recently entered my life. Kat Toups and Nancy Heydemann have been incredible friends, and I can't thank them enough for their support. Thank you to the advisers and healers who kept me grounded and on-track over the past two years: Dawna Ara, Linda Backman, Lisa Barnett, Keith Bailey, Ulrich Bold, Jennifer Brinn, Carl Buchheit, Janet Co, Andrew Cohen, Karen Curry, Michael Fishman, Richard Handy, Eva Herr, Denmo Ibrahim, Sabine Kaiser, Elanita Korian, Adam Markel, Allison Post, Laura Powers, Elsa Sunita, Edith Ubuntu Chan, Kimberly Urrea, Michelle Veneziano, and Catherine Yunt.

I thank my dear friend Cameron Goldberg for giving me the initial encouragement to pursue my passions around these topics when I was at first skittish. And I thank my friends Tray McCurdy and Holt Mettam for nudging me to write this book at a dinner in the summer of 2017. So many wonderful friends have been particularly supportive during my process of writing this book over the last year: Jon Brick, Cameron Goldberg, Nick Marshall, Scott Eisen, Drew Singleton, Ross Exler, Stephen Garten, Eric White, Andrew Wilen, Brad Nelson, Shalin Parikh, Tray McCurdy, Casey McCourt, Nate DeOms, Neil Vangala, Evan Daar, Gideon Lowin, Santi Nuñez, Varun Gehani, Bryant Yung, Alexis Sclamberg, Ross Barasch, Shailesh Sachdeva, Jen Leybovich, Tiffany Young, Blake Brinker, Gaurav Sharma, Gordon Green, Greg Miller, Sangeet Sood, Spencer Ton, Alyssa Smilowitz, Scott DeBenedett, Phil Farinacci, Kane Hochster, Amelia Salyers, Julia Peppiatt, Jonah Wagner, John Snyder, Sara Peters, Parker Preyer, Matt Spindler, Alex Faust, Vittal Kadapakkam, Ilya Trubov, Charlie Brosens, Hans Plukas, Andrew Doupé, David Levit, Jason Smith, Jackie Knechtel, Justin Faerman, Lars King, Chirag Kulkarni, Josh Fields, Stephen Sokoler, Rochel Leah Bernstein, Lilian Wang, Matt Hellauer, Rachel Braver, Justin Steinfelder, Tyler Heishman, Liza Connolly, Lara Avsar, Kevin Waldman, Natalie Carlstead, Nadine Kedrus Marshall, Jackie Wolfson, Nora Nagle, Emily Polidan, Bailey Gerber, Jessica Theroux, Sibel Yalman, David Hopkins, Sophia Fleischer, Leonid Rozkin, Adam Spector, Mikhail Simonov, Walter Barker, Jessica Aycock, Parth Shah, Jon Stein, and Ellyn Guttman. There are many other friends who have been supportive throughout different periods of my life whom I thank as well. I also thank my longtime friend and podcast producer Matt Ford for his support.

I am grateful to have had such an excellent education, where I learned research skills and how to express ideas in writing. Thank you to McDonogh School in Baltimore, Maryland, and Princeton University.

My many years competing as a tennis player undoubtedly contributed to my being able to write this book. I learned the value of perseverance, hard work, integrity, and humility through the countless hours spent with my coaches: George Martin IV, Laddie Levy (McDonogh), and Glenn Michibata (Princeton). I also thank my teammates on the McDonogh and Princeton tennis teams.

I thank my business colleagues and clients whom I've worked with over the past ten years. The skills developed in communication and narrative-building unquestionably assisted my writing of this book. A special thanks to my colleagues Kevin Rivette, Ralph Eckardt, Peter Detkin, and Andy Filler for their mentoring and guidance in my professional development.

None of this would have been possible without the unrelenting support and love from my parents, Bob and Karen, and my brothers, Zack and Jake. Thank you to family members who are no longer with us, but who played a huge role in shaping the person I am today: Fred and Sandy Hittman, Rosie, Tara, and Susie.

Finally, I thank the brave scientists whose research enabled me to write this book.

Glossary

After-death communications: Incidents in which living individuals claim to receive signs or communications from deceased individuals.

Butterfly effect: The notion in chaos theory that small changes in initial conditions can have an enormous, nonlinear effect on outcomes (e.g., a butterfly flapping its wings in China can eventually cause a hurricane in New York).

Consciousness: The subjective sense of being aware and experiencing life. When you say, "I am reading these words," "I" is the consciousness having the experience.

Deathbed vision: An experience shortly before death in which one is visited by the deceased (e.g., a deceased relative). Sometimes people report feelings of peace and unconditional love, not dissimilar from what is described in near-death experiences.

Discarnate: A deceased individual with whom a psychic medium is attempting to communicate.

Effect size: The size of a statistical effect indicating the extent of departure from randomness.

Energy healing: The ability to heal another using one's mind.

Entanglement: The generally accepted notion in quantum physics that particles physically distant from one another have hidden connections that cannot be seen with our eyes; Albert Einstein famously called it "spooky action at a distance."

Fear-death experience: An experience resembling a near-death experience, wherein the individual is not sick or physically harmed but merely thinks that he or she will die.

File-drawer accusation: An accusation waged against an experimenter, claiming the experimenter's positive results were only positive because he or she simply hid negative results.

Ganzfeld experiments: Telepathy studies in which one participant sends mental images to another subject using a specific procedure.

Global Consciousness Project: A continuation of research conducted at the Princeton Engineer Anomalies Research Lab (PEAR), which statistically examines the behavior of 1's and 0's in random number generators around the world during major world events.

Life review: An experience reported in many near-death experiences in which individuals review their entire lives in an instant, reliving old events both from their perspectives *and* from that of those they affected during their lives.

Materialism: The framework on which most of modern science is built, which says that the universe is fundamentally made of matter. It says that matter (the brain) creates consciousness.

Maternal impressions: Instances in which a pregnant woman sees a distributing image (such as a physical deformity), and her baby is then born with the deformity that she saw while pregnant.

Mediums (also referred to as psychic mediums): Living individuals who can communicate with the deceased.

Near-death experience (NDE): A transformative, hyper-real experience reported by people who are nearing death, typically after extreme trauma (e.g., drowning, brain damage, cardiac arrest).

Newtonian physics: A branch of physics that conforms to our everyday experience, but only serves as an approximation for reality.

Nonlocal consciousness: The idea that consciousness is not localized, or confined to, the body/brain.

Precognition (presentiment): The ability to accurately know or feel something before it happens; knowing the future when the future is ostensibly unknown.

Precognitive dreams: Dreams that accurately predict future events.

Princeton Engineering Anomalies Research Lab (PEAR): A lab at Princeton University that existed from 1979 to 2007 and was run by Princeton's former dean of engineering, Dr. Robert Jahn, and laboratory manager Brenda Dunne. PEAR studied anomalies suggestive of nonlocal consciousness (e.g., remote viewing, psychokinesis, etc.).

Prospective cardiac arrest studies: A type of near-death experience study in which survivors of cardiac arrest are interviewed about the experiences they had while they were in cardiac arrest.

Psychokinesis: The ability of the mind to have a physical impact on matter.

Quantum physics (quantum mechanics): A branch of physics conceived in the early 1900s that studies small particles. The findings are mind-boggling and suggest that reality works in ways that defy common sense.

Reincarnation: The idea that one's consciousness has inhabited the body of another individual. Sometimes memories from the "previous life" are remembered in the current life (i.e., 50 years of studies at the University of Virginia on 2,500-plus children who recall previous lives).

Savants: Individuals who possess remarkable mental abilities but also have severe impairments.

Shared-death experience: When a healthy bystander has an NDE-like experience around a person who is dying.

Spoon-bending: The ability to bend a metal spoon using one's mind alone.

Stargate Project: A U.S. government-sponsored program to study psychic phenomena such as remote viewing (~1972–1995).

Telepathy: Mind-to-mind communication.

Telesomatic events: Experiences in which one person physically feels the same thing or experiences the same emotion that another feels. It is often reported in identical twins.

Terminal lucidity: Unexplained, sudden mental clarity reported in patients who are approaching death but had previously suffered severe cognitive impairments (e.g., Alzheimer's disease).

Time dilation: The concept in Einstein's general relativity that time moves more slowly when one is moving at a high velocity and/or is in a high-gravity environment (relative to someone moving slower or in a low-gravity environment).

Transcendental awareness: An ability to perceive (e.g., "see," "hear," etc.) even when one's organs are ostensibly nonfunctional. The awareness seems to operate independently of bodily function and might explain how some blind individuals report being able to "see" during their NDEs.

Veridical out-of-body experience: When an individual having an NDE later describes hovering over his or her body and saw or heard things that are verified as being accurate by others in the room.

Windbridge Research Center: An organization that studies death and dying, including controlled studies on psychic mediums.

Xenoglossy: Speaking a language that one was never taught.

Endnotes

Preface

1 Dalio, *Principles*, 268.
2 Figures A, B, and C are adapted from Dr. Dean Radin (e.g., *Real Magic*, page 197).
3 A critique of this argument that I've heard is as follows: "It is untenable to maintain that there is no reality independent of consciousness, for there is plenty of evidence about what was going on in the universe before consciousness evolved." However, as Dr. Kastrup points out, this argument is flawed because it assumes materialism in its very defense of materialism. It assumes that consciousness evolves through biology, which is the very subject in question. In Dr. Kastrup's words, the argument suffers because it "assumes materialism—the notion that consciousness is generated by, and confined to, biological nervous systems—in a circular argument for materialism. If all reality is in consciousness itself, then it is nervous systems that are in consciousness, not consciousness in nervous systems." See *Brief Peeks Beyond* by Dr. Bernardo Kastrup, 24-25.
4 Spira, *The Nature of Consciousness*, 149.
5 Kastrup, *Why Materialism Is Baloney*, 63.
6 Ibid., 204.
7 King, Interview: *Richard Dawkins Celebrates Reason, Ridicules Faith,* https://www.npr.org/sections/13.7/2012/03/26/149310560/atheist-firebrand-richard-dawkins-unrepentant-for-harsh-words-targeting-faith.
8 Gosling, *Science and the Indian Tradition: When Einstein Met Tagore*, 162.
9 Russell, *From Science to God*, 28.
10 Internet Encyclopedia of Philosophy, *The Hard Problem of Consciousness.*
11 *Science* magazine's 125th-anniversary issue, http://www.sciencemag.org/site/feature/misc/webfeat/125th/.
12 Harris, *Waking Up*, 60.
13 As cited in Russell, *From Science to God*, 26.
14 Kastrup, *Why Materialism Is Baloney*, 31.
15 Interview with Max Planck, *The Observer.*
16 Jeans, *The Mysterious Universe.*
17 Schiller, *Riddles of the Sphinx: A Study in the Philosophy of Humanism.*
18 Kastrup, *Why Materialism Is Baloney*, 81.
19 *Science* magazine's 125th-anniversary issue, http://www.sciencemag.org/site/feature/misc/webfeat/125th/.

Chapter 1

1 As cited in Powell, *The ESP Enigma*, 229.

2 "The brain as an antenna" analogy is not fully precise. It suggests that the brain and consciousness are separate. On the contrary, this book makes the case that a single underlying consciousness is the fundamental medium of reality, which implies that brains are simply localization processes of consciousness (for more on this topic, see *Brief Peeks Beyond* by Dr. Bernardo Kastrup (2015), page 163). I reference the antenna analogy in this book because it is at least a departure from the materialist view of consciousness and is easy to grasp for readers who are new to these concepts. The analogy is effective in that it conveys the idea that consciousness does not originate from the brain. It can be viewed as a "gateway" analogy on the path to a more complete and precise theory.

3 Exemplary podcasts include: Healing Powers (particularly the early episodes), The Cosmos in You, Skeptiko, Provocative Enlightenment, Extreme Health Radio, Inspire Nation, and many others.

4 Spira, *The Nature of Consciousness*, 149.

5 Schrödinger, *What Is Life? with Mind and Matter*, 139.

6 "Telepathy," *Wikipedia*, last modified April 22, 2018.

7 Closer to Truth. "Lawrence Krauss—Does ESP Make Sense?" YouTube video, 8:30. Posted July 17, 2017. https://youtu.be/5NweHLQmbZE.

8 For examples of scientific papers, see "Selected Psi Research Publications," made available for free by Dr. Dean Radin at http://deanradin.com/evidence/evidence.htm.

9 Utts, *From Psychic Claims to Science: Testing Psychic Phenomena with Statistics.*

10 Schwartz, "Through Time and Space: The Evidence for Remote Viewing" in *Evidence for Psi: Thirteen Empirical Research Reports*, ed. Broderick and Goertzel.

11 McKie, *Royal Mail's Nobel guru in telepathy row*, https://www.theguardian.com/uk/2001/sep/30/robinmckie.theobserver.

12 Smolin, *Time reborn: From the crisis in physics to the future of the universe.*

13 Alexander, *Proof of Heaven*, 151.

14 Radin, http://www.deanradin.com/NewWeb/bio.html.

15 Radin, *Real Magic*, 97.

16 Radin, *Supernormal, Entangled* Minds, and *The Conscious* Universe. Also see Tressoldi. *Extraordinary claims require extraordinary evidence: The case of non-local perception, a classical and Bayesian review of evidences.*

17 Tressoldi. *Extraordinary claims require extraordinary evidence: The case of non-local perception, a classical and Bayesian review of evidences.* Also, Williams, *Revisiting the ganzfeld ESP debate: A basic review and assessment.*

18 Mossbridge, Tressoldi, & Utts, *Predictive physiological anticipation preceding seemingly unpredictable stimuli: A meta-analysis.* Also Bem, Tressoldi, Raberyon, & Duggan. *Feeling the future: A meta-analysis of 90 experiments on the anomalous anticipation of random future events.*

19 Bosch, Steinkamp, & Boller. *Examining psychokinesis: The interaction of human intention with random number generators—a meta-analysis.* Radin, Nelson, Dobyns, & Houtkooper, *Re-examining psychokinesis: Commentary on the Bösch, Steinkamp and Boller meta-analysis.* Nelson, Radin, Shoup, & Bancel. *Correlations of continuous random data with major world events.* Also see global-mind.org/results.html.

20 Radin, *Real Magic*, 97.
21 Cardeña, *The Experimental Evidence for Parapsychological Phenomena: A Review.*
22 Utts, *An Assessment of Evidence for Psychic Functioning.*
23 Coyne, *Science Is Being Bashed by Academics Who Should Know Better*, https://newrepublic.com/article/117244/jeffrey-kripals-anti-materialist-argument-promotes-esp.
24 Kelly, 'Introduction: Science and Spirituality at a Crossroads' in *Beyond Physicalism*, Kelly et al., xv.
25 Carroll, *The Big Picture*, 154.
26 McKie, *Royal Mail's Nobel guru in telepathy row*, https://www.theguardian.com/uk/2001/sep/30/robinmckie.theobserver.
27 Pinker, Praise for Carroll's *The Big Picture.*
28 Parker and Brusewitz, *A Compendium of the Evidence for Psi.*
29 Closer to Truth. "Lawrence Krauss—Does ESP Make Sense?" YouTube video, 8:30. Posted July 17, 2017. https://youtu.be/5NweHLQmbZE.
30 Jahn and Dunne, *Consciousness and the Source of Reality*, 32.
31 Hofstadter, *A Cutoff for Craziness*, https://www.nytimes.com/roomfordebate/2011/01/06/the-esp-study-when-science-goes-psychic/a-cutoff-for-craziness.
32 Turing, *Computing Machinery and Intelligence.*
33 Sheldrake, *Dogs That Know When Their Owners Are Coming Home*, 330.
34 Schnabel, *Remote Viewers*, 7.
35 Interview with Ray Hyman, *Austin American-Statesman*, July 14, 2002.
36 Eysenck, *Sense and Nonsense in Psychology.*
37 Rosenblum and Kuttner, *Quantum Enigma*, 255.
38 Lilou Mace. "Psychoenergetics—William Tiller Ph.D." YouTube video, 43:41. Posted October 15, 2012. https://www.youtube.com/watch?v=pI7jO1JuF-c&app=desktop.
39 Barušs and Mossbridge, *Transcendent Mind*, 21.
40 Sommer, *Psychical research in the history and philosophy of science. An introduction and review.*
41 Tiller, *Psychoenergetic Science: A Copernican Scale Revolution.*
42 As cited in Swanson, *The Synchronized Universe: New Science of the Paranormal*, 4.
43 As cited in Radin, *Entangled Minds*, 211.
44 Greene, *The Fabric of the Cosmos*, 9.
45 Utts, *An Assessment of Evidence for Psychic Functioning.*
46 Ibid.
47 Tyson, Neil deGrasse. Twitter Post. June 14, 2013, 7:41am, https://twitter.com/neiltyson/status/345551599382446081?lang=en.
48 Utts, *Appreciating Statistics.*
49 Schwartz, foreword to *The Reality of ESP*, xv.
50 Josephson, *Pathological Disbelief.*
51 Sheldrake, *Dogs That Know When Their Owners Are Coming Home*, 331.
52 Ibid., 330.
53 Knox, *Science, God and the Nature of Reality: Bias in Biomedical Research.*
54 Sagan, *The Dragons of Eden*, 7.
55 Schwartz, *Opening to the Infinite*, 16.
56 Sagan, *The Demon-Haunted World*, 302.

57 Sheldrake, *Sir John Maddox—Book for Burning*, https://www.sheldrake.org/reactions/sir-john-maddox-book-for-burning.
58 Baruš and Mossbridge, *Transcendent Mind*, 25.
59 Mossbridge, 'Physiological Activity That Seems to Anticipate Future Events' in *Evidence for Psi: Thirteen Empirical Research Reports*, ed. Damien Broderick and Ben Goertzel, 58.
60 For example, as cited in Collective Evolution, *Reality is, as Tesla said, 'non-physical.' Why the large hadron collider will never find the 'god particle'*, http://www.collective-evolution.com/2016/11/15/most-of-reality-is-as-tesla-said-non-physical-why-the-large-hadron-collider-will-never-find-the-god-particle/.
61 Jahn and Dunne, *Consciousness and the Source of Reality*, 337.
62 Tiller, http://www.tillerinstitute.com/.

Chapter 2

1 Harari, *Homo Deus*, 108-109.
2 Harris, *Waking Up*, 60.
3 Friedman, *A neurosurgeon calls this basic fact about the brain 'too strange to understand,'* http://www.businessinsider.com/the-strangest-thing-about-the-brain-2016-6.
4 Tsakiris, *Why Science Is Wrong*, 7-8.
5 Lanza and Berman, *Biocentrism*, 4.
6 If consciousness is indeed the fundamental medium of reality, as this book suggests, then we might be striving for the impossible by trying to define it using language. Language is inherently limiting. Consciousness might be limitless, even across space and time. And if that is the case, any word used to describe it will miss its unlimited essence.
7 Russell, *From Science to God*, 83-84.
8 Sheldrake, *Science Set Free*, 10.
9 Noë, *Out of Our Heads*, xi.
10 Sheldrake, *Science Set Free*, 9.
11 Dossey, *One Mind*, 81-82.
12 Schwartz, *The Afterlife Experiments*, 267.
13 Powell, *The ESP Enigma*, 23.
14 Suzannah Scully (Cosmos in You), interview with Dr. Diane Powell, May 9, 2017, https://www.suzannahscully.com/full-episodes.
15 Alexander, 'Near-Death Experiences: The Mind-Body Debate & the Nature of Reality' in the *Science of Near-Death Experiences*, ed. John C. Hagan III, 108-109.
16 Burt, *ESP and Psychology*.
17 *Manifesto for a Post-Materialist Science*, http://opensciences.org/about/manifesto-for-a-post-materialist-science.
18 For more on this general topic, see Bernardo Kastrup's March 2017 article entitled *Transcending the Brain* available at https://blogs.scientificamerican.com/guest-blog/transcending-the-brain/.
19 Costandi, *Psychedelic chemical subdues brain activity*, https://www.nature.com/news/psychedelic-chemical-subdues-brain-activity-1.9878.
20 Carhart-Harris et al., *Neural correlates of the psychedelic state as determined by fMRI studies with psilocybin.*
21 Ibid.

22 Kelly and Presti, 'A Psychobiological Perspective on 'Transmission Models" in *Beyond Physicalism*, Kelly et al., 143.
23 E.W. Kelly et al., 'Unusual Experiences Near Death and Related Phenomena' in *Irreducible Mind*, Kelly & Kelly et al., 386.
24 Nahm et al., *Terminal lucidity: A review and a case collection.*
25 Ibid.
26 Barušs and Mossbridge, *Transcendent Mind*, 104.
27 *Kim Peek—The Real Rain Man*, https://www.wisconsinmedicalsociety.org/professional/savant-syndrome/profiles-and-videos/profiles/kim-peek-the-real-rain-man/.
28 Suzannah Scully (Cosmos in You), interview with Dr. Diane Powell, May 9, 2017, https://www.suzannahscully.com/full-episodes.
29 Treffert, *Islands of Genius*, 122.
30 Ibid., 123.
31 Ibid., 25.
32 Kelly, 'Empirical Challenges to Theory Construction' in *Beyond Physicalism*, Kelly et al., 16-17.
33 Suzannah Scully (Cosmos in You), interview with Dr. Diane Powell, May 9, 2017, https://www.suzannahscully.com/full-episodes.
34 Dossey, *One Mind*, 132.
35 Treffert, *Islands of Genius*, xv.
36 Ibid., xiv.
37 Eldon Taylor (Provocative Enlightenment), interview with Darold Treffert, MD, August 9, 2017, http://www.provocativeenlightenment.com/.
38 McTaggart, *The Field*, 78.
39 Ibid.
40 As cited in Moody, *Glimpses of Eternity*, 59.
41 Pietsch, *Shufflebrain*, 213.
42 Ibid., 39-40.
43 Talbot, *The Holographic Universe*, 26.
44 McTaggart, *The Field*, 86.
45 Pietsch, *Shufflebrain*, 92-93.
46 Ibid., 98.
47 Ibid., 1.
48 Pearsall, *The Heart's Code*, 7.
49 Ibid., 90.
50 Van Lommel, *Consciousness beyond Life*, 282.

Chapter 3

1 Heisenberg, *Physics and Philosophy: The Revolution in Modern Science*, 80-81.
2 *Experiment confirms quantum theory weirdness*, https://phys.org/news/2015-05-quantum-theory-weirdness.html.
3 Jauch, *The Problem of Measurement in Quantum Mechanics.*
4 J.A. Wheeler in J. Mehra (ed.), *The Physicist's Conception of Nature*, 244.
5 Rosenblum and Kuttner, *Quantum Enigma*, 95.
6 Feynman, *The character of physical law.*
7 Currivan, *The Cosmic Hologram*, 53.
8 Hawking and Mlodinow, *The Grand Design*, 7.

9 Figure reconstructed from NASA introduction to electromagnetic spectrum https://science.nasa.gov/ems/01_intro.
10 Kelly and Presti, 'A Psychobiological Perspective on 'Transmission Models" in *Beyond Physicalism*, Kelly et al., 123.
11 Suzannah Scully (Cosmos in You), interview with Donald Hoffman, July 27, 2015, https://www.suzannahscully.com/full-episodes.
12 Rosenblum and Kuttner, *Quantum Enigma*, 269.
13 Ibid., 54.
14 Greene, *The Fabric of the Cosmos*, 199.
15 Hawking and Mlodinow, *The Grand Design*, 67.
16 Lowery, *Study showing that humans have some psychic powers caps Daryl Bem's career*, http://news.cornell.edu/stories/2010/12/study-looks-brains-ability-see-future.
17 E.F. Kelly, 'Toward a Psychology for the 21st Century' in *Irreducible Mind*, Kelly & Kelly et al., 611.
18 Currivan, *The Cosmic Hologram*, 52.
19 Albert Einstein, letter to Max Born, *The Born-Einstein Letters*, quoted variously, including Andrew Robinson *Einstein: A Hundred Years of Relativity*, 82.
20 Einstein et al., *Can Quantum-Mechanical Description of Physical Reality Be Considered Complete?*
21 Bell, *On the Einstein Podolsky Rosen Paradox*.
22 Markoff, *Sorry, Einstein. Quantum Study Suggests 'Spooky Action' Is Real.*, https://www.nytimes.com/2015/10/22/science/quantum-theory-experiment-said-to-prove-spooky-interactions.html.
23 Rosenblum and Kuttner, *Quantum Enigma*, 188.
24 Bohm and Hiley, *One the Intuitive Understanding of Nonlocality as Implied by Quantum Theory*.
25 Radin, *Entangled Minds*, 19.
26 Hawking and Mlodinow, *The Grand Design*, 68.
27 Feynman, *Quantum Behavior*, http://www.feynmanlectures.caltech.edu/III_01.html#Ch1-S7.
28 Ibid., 73.
29 Rosenblum and Kuttner, *Quantum Enigma*, 83.
30 Greene, *The Hidden Reality*, 201.
31 As cited in Lanza and Berman, *Beyond Biocentrism*, 61.
32 As cited in Lanza and Berman, *Biocentrism*, 181.
33 Paulson, *Roger Penrose On Why Consciousness Does Not Compute*, http://nautil.us/issue/47/consciousness/roger-penrose-on-why-consciousness-does-not-compute.
34 Dan Jacob. "Neil deGrasse on consciousness." YouTube video, 1:14. Posted June 26, 2015. https://www.youtube.com/watch?v=QGekFhbyQLk.
35 Carr, 'Hyperspatial Models of Matter and Mind' in *Beyond Physicalism*, Kelly et al., 228.
36 Wigner, *Symmetries and Reflections*.
37 Wigner, 'Remarks on the Mind-Body Problem,' in *The Scientist Speculates*, (ed.) Good, 289.
38 Wigner, 'The Place of Consciousness in Modern Physics' in *Philosophical Reflections and Syntheses*, Wigner, 263.
39 Herr, *Consciousness*, 61.

40 Radin et al., *Psychophysical interactions with a single-photon double-slit optical system* and Stapp, *Quantum Theory and Free Will.*
41 Penrose and Hameroff, *Consciousness in the Universe: Neuroscience, Quantum Space-Time Geometry and Orch OR Theory.*
42 Interview with Max Planck, The Observer.
43 Ananthaswamy, *A classic quantum test could reveal the limits of the human mind*, https://www.newscientist.com/article/2131874-a-classic-quantum-test-could-reveal-the-limits-of-the-human-mind/.
44 Ibid.
45 The results of the studies are described in a talk by Dr. Dean Radin given at the 2016 The Science of Consciousness conference: https://www.youtube.com/watch?v=nRSBaq3vAeY.
46 Radin et al., *Consciousness and the double-slit interference pattern: Six experiments* and Radin et al., *Psychophysical interactions with a double-slit interference pattern.*
47 Radin et al., *Psychophysical interactions with a single-photon double-slit optical system.*
48 Radin, Dean. Twitter Post. December 15, 2015, 12:42pm, https://twitter.com/deanradin/status/676865032009814016.
49 The Institute of Noetic Sciences : IONS. "New Experiments Show Consciousness Affects Matter ~ Dean Radin Ph.D." YouTube video, 41:04. Posted June 7, 2016. https://www.youtube.com/watch?v=nRSBaq3vAeY.
50 Guerrer, *Consciousness-Related Interactions in a Double-Slit Optical System.*
51 Powell, *The ESP Enigma*, 180.
52 As cited in Russell, *From Science to God*, 48.
53 Ibid., 49.
54 As cited in Ibid.
55 Capra, *The Tao of Physics*, 203.
56 Ibid., 140.
57 Ibid., 140-141.
58 Barušs and Mossbridge, *Transcendent Mind*, 8.
59 Capra, *The Tao of Physics*, 133.
60 Hawking and Mlodinow, *The Grand Design*, 98.
61 Ibid., 99. Also see Hafele and Keating, *Around-the-World Atomic Clocks: Predicted Relativistic Time Gains.*
62 Rosenblum and Kuttner, *Quantum Enigma*, 96.
63 *Experiment confirms quantum theory weirdness*, https://phys.org/news/2015-05-quantum-theory-weirdness.html.
64 Wheeler, 'Law without Law' in *Quantum Theory and Measurement*, Wheeler, Zurek, (Eds.), 184.
65 Buchheit, *Transformational NLP*, 120.
66 As cited in Powell, *The ESP Enigma*, 196-197.
67 *The relativity of space and time*, http://www.einstein-online.info/elementary/specialRT/relativity_space_time.html.
68 Greene, *The Fabric of the Cosmos*, 8
69 Gleik, *Chaos*, 3.
70 Ibid., 6.
71 Powell, *The ESP Enigma*, 186.
72 Gleik, *Chaos*, 8.
73 Hawkins, *Power vs. Force*, 60.

Chapter 4

1 Schnabel, *Remote Viewers*, 215.
2 Targ, *The Reality of ESP*, 4.
3 Schnabel, *Remote Viewers*, 29.
4 Schnabel, *Remote Viewers*, back cover.
5 Targ and Harary, *The Mind Race*.
6 Schnabel, *Remote Viewers*, back cover.
7 Penman, *Could there be proof to the theory that we're ALL psychic?*, http://www.dailymail.co.uk/news/article-510762/Could-proof-theory-ALL-psychic.html.
8 Swanson, *The Synchronized Universe: New Science of the Paranormal*, 10.
9 "Stargate Project," Wikipedia, last modified April 28, 2018.
10 Targ, *The Reality of ESP*, 44.
11 Schwartz, "Through Time and Space: The Evidence for Remote Viewing" in *Evidence for Psi: Thirteen Empirical Research Reports*, ed. Broderick and Goertzel.
12 Targ, *The Reality of ESP*, 19.
13 Ibid., 22.
14 Ibid., 32.
15 Ibid., 31-32.
16 For the drawing itself, see *The Reality of ESP*, 33.
17 Ibid., 32.
18 Ibid., 116.
19 Dossey, *One Mind*, 159.
20 Skeptiko, *Ex-Stargate Head, Ed May, Unyielding Re Materialism, Slams Dean Radin |341|*, podcast 341 http://skeptiko.com/ed-may-unyielding-re-materialism-341/.
21 Targ, *The Reality of ESP*, 15.
22 Radin, *Entangled Minds*, 292.
23 Allen, *How Uri Geller convinced the CIA he was a 'psychic warrior'*, http://www.telegraph.co.uk/news/2017/01/18/uri-geller-convinced-cia-psychic-warrior/.
24 CIA-RDP79-00999A000300030027-0.
25 Ibid.
26 Targ, *The Reality of ESP*, 164.
27 CIA-RDP96-00789R002100220001.
28 *Homepage for Professor Jessica Utts*, http://www.ics.uci.edu/~jutts/.
29 Utts, *An Assessment of the Evidence for Psychic Functioning*.
30 Ibid.
31 Utts, *From Psychic Claims to Science: Testing Psychic Phenomena with Statistics*.
32 Hyman, *Evaluation of Program on Anomalous Mental Phenomena*, http://www.ics.uci.edu/~jutts/hyman.html.
33 Schwartz, Foreword to *The Reality of ESP*, xiii.
34 Radin, *Supernormal*, 142.
35 Jahn and Dunne, *Consciousness and the Nature of Reality*, 235.
36 Ibid., 234.
37 Radin, *Supernormal*, 142.
38 Penman, *Could there be proof to the theory that we're ALL psychic?*, http://www.dailymail.co.uk/news/article-510762/Could-proof-theory-ALL-psychic.html.
39 Dossey, *One Mind*, 164.

Chapter 5

1 Radin, *Real Magic*, 98.
2 Bem and Honorton, *Does psi exist? Replicable evidence for an anomalous process of information transfer.*
3 Turing, *Computing Machinery and Intelligence.*
4 Eldon Taylor (Provocative Enlightenment), interview with Darold Treffert, MD, August 9, 2017, http://www.provocativeenlightenment.com/.
5 McKie, *Royal Mail's Nobel guru in telepathy row*, https://www.theguardian.com/uk/2001/sep/30/robinmckie.theobserver.
6 Kaku, *The Future of Mind*, 64.
7 Radin, *Supernormal*, 137.
8 Skeptiko, *Ex-Stargate Head, Ed May, Unyielding Re Materialism, Slams Dean Radin |341|*, podcast 341 http://skeptiko.com/ed-may-unyielding-re-materialism-341/.
9 Rosenblum and Kuttner, *Quantum Enigma*, 255.
10 "Ganzfeld experiment," Wikipedia, last modified March 21, 2018.
11 Radin, *Entangled Minds*, 118.
12 Bem and Honorton, *Does psi exist? Replicable evidence for an anomalous process of information transfer.*
13 Radin, *Supernormal*, 190. For more on ganzfeld experiments, see Williams, B.J. (2011). Revisiting the ganzfeld ESP debate: A basic review and assessment. *Journal of Scientific Exploration*, 25(4), 639-661.
14 Radin, *Entangled Minds*, 121.
15 Barušs and Mossbridge, *Transcendent Mind*, 42.
16 Sheldrake, *Dogs That Know When Their Owners Are Coming Home*, 330.
17 Sheldrake, *The Sense of Being Stared At*, 162.
18 Ibid., 163.
19 Ibid.
20 Sheldrake, *Science Set Free*, 222.
21 Ibid., 222-224.
22 Ibid., 224.
23 Radin, *Entangled Minds*, 127.
24 Sheldrake, *Science Set Free*, 244.
25 Ibid., 245.
26 Ibid., 246.
27 Sheldrake, *The Sense of Being Stared At: Part I: Is it Real or Illusory?*
28 Playfair, *Twin Telepathy*, 151.
29 Ibid.
30 Ibid.
31 Ibid., 139.
32 Ibid., 140.
33 Ibid., 16.
34 Dossey, Unbroken Wholeness: The Emerging View of Human Interconnection, http://realitysandwich.com/170309/human_interconnection/.
35 Playfair, *Twin Telepathy*, 148.
36 Ibid., 143.
37 Ibid., 145.

38 Dossey, Unbroken Wholeness: The Emerging View of Human Interconnection, http://realitysandwich.com/170309/human_interconnection/.
39 Powell et al., *Non Local Consciousness in an Autistic Child*, http://www.consciousness.arizona.edu/documents/TSC2017AbstractBook-final-5.10.17-final.pdf.
40 Eldon Taylor (Provocative Enlightenment), interview with Darold Treffert, MD, August 9, 2017, http://www.provocativeenlightenment.com/.
41 Dossey, *One Mind*, 133.
42 Suzannah Scully (Cosmos in You), interview with Dr. Diane Powell, May 9, 2017, https://www.suzannahscully.com/full-episodes.

Chapter 6

1 Utts, *An Assessment of Evidence for Psychic Functioning.*
2 Radin, *Supernormal*, 177.
3 Kelly, 'Introduction: Science and Spirituality at a Crossroads' in *Beyond Physicalism*, Kelly et al., 6.
4 MacIsaac, *Neuroscientist discusses precognition—or 'mental time travel'*, https://www.theepochtimes.com/uplift/neuroscientist-discusses-precognition-or-mental-time-travel_2362702.html.
5 Radin, *The Conscious Universe*, 98.
6 As cited in Radin, *Entangled Minds*, 69.
7 Radin, *The Conscious Universe*, 100.
8 Radin, *Supernormal*, 133.
9 Radin, *Supernormal*, 134. Also see Honorton and Ferrari, *"Future telling": A meta-analysis of forced-choice precognition experiments.*
10 Radin, *Supernormal*, 135.
11 Radin, *Entangled Minds*, 165.
12 Ibid., 168.
13 Ibid., 170.
14 Radin, *Supernormal*, 151.
15 Radin, *Entangled Minds*, 179.
16 Radin, *Supernormal*, 154-155.
17 Radin, *Entangled Minds*, 172.
18 Radin, *Supernormal*, 167.
19 Lowery, *Study showing that humans have some psychic powers caps Daryl Bem's career*, http://news.cornell.edu/stories/2010/12/study-looks-brains-ability-see-future.
20 Carey, *Journal's Paper on ESP Expected to Prompt Outrage*, http://www.nytimes.com/2011/01/06/science/06esp.html.
21 Bem, *Feeling the Future.*
22 Carey, *Journal's Paper on ESP Expected to Prompt Outrage*, http://www.nytimes.com/2011/01/06/science/06esp.html.
23 Hofstadter, *A Cutoff for Craziness*, https://www.nytimes.com/roomfordebate/2011/01/06/the-esp-study-when-science-goes-psychic/a-cutoff-for-craziness.
24 Carey, *Journal's Paper on ESP Expected to Prompt Outrage*, http://www.nytimes.com/2011/01/06/science/06esp.html.
25 Bem et al., *Feeling the future: A meta-analysis of 90 experiments on the anomalous anticipation of random future events*, https://f1000research.com/articles/4-1188/v1.

26 Mossbridge et al., *Predictive physiological anticipation preceding seemingly unpredictable stimuli: a meta-analysis.*

27 "Julia Mossbridge," http://noetic.org/directory/person/julia-mossbridge.

28 Dossey, *The Power of Premonitions*, xvii-xviii.

29 *The Power of Premonitions: How Knowing the Future Can Shape Our Lives, An Interview with Larry Dossey*, http://www.dosseydossey.com/larry/Interview_Questions-Premonitions.pdf. Dr. Dossey recalls the account from Sally Rhine Feather and Michael Schmicker. *The Gift: ESP, the Extraordinary Experiences of Ordinary People*. New York: St. Martin's. 2005, pages 1-3.

30 Mazza, *Virginia Man Wins The Lottery By Playing Numbers From A Dream*, https://www.huffingtonpost.com/entry/lottery-dream-numbers-jackpot_us_5a79402be4b00f94fe9456f0.

31 Watt, *Precognitive dreaming: Investigating anomalous cognition and psychological factors.*

32 Barušs and Mossbridge, *Transcendent* Mind, 67.

33 Powell, *The ESP Enigma*, 78.

34 Talbot, *The Holographic Universe*, 216.

35 Utts, *An Assessment of Evidence for Psychic Functioning.*

Chapter 7

1 Sheldrake, *Dogs That Know When Their Owners are Coming Home*, 2.

2 InfinicityFilm. "Rupert Sheldrake—Dogs Who Know When Their Owners Are Coming Home." YouTube video, 1:41. Posted April 22, 2012. https://www.youtube.com/watch?v=9QsPWitQovM.

3 Radin, *Supernormal*, 82.

4 *Jaytee—A Dog Who Knew When His Owner Was Coming Home: The ORF Experiment* , https://www.sheldrake.org/videos/jaytee-a-dog-who-knew-when-his-owner-was-coming-home-the-orf-experiment.

5 Sheldrake and Smart, *A Dog That Seems to Know When His Owner Is Coming Home: Videotaped Experiments and Observations*, https://www.sheldrake.org/research/animal-powers/a-dog-that-seems-to-know-when-his-owner-is-coming-home-videotaped-experiments-and-observations.

6 Sheldrake, *Dogs That Know When Their Owners Are Coming Home*, 63.

7 Ibid., 317.

8 Radin, *Supernormal*, 83.

9 Sheldrake, *Dogs That Know When Their Owners Are Coming Home*, 176.

10 Ibid., 239.

11 Ibid., 238-239.

12 Ibid., 267.

13 Ibid.

14 Ibid., 277-283.

15 Ibid., 107.

16 For example, Peoc'h, *Psychokinetic Action of Young Chicks on the Path of An Illuminated Source.*

17 Sheldrake, *Dogs That Know When Their Owners Are Coming Home*, 296-297.

Chapter 8

1 Jahn and Dunne, *Margins of Reality*, 144.
2 Radin, *Real Magic*, 235.
3 Jacobsen, *Phenomena*, 403.
4 Tiller, http://www.tillerinstitute.com/.
5 Swanson, *The Synchronized Universe: New Science of the Paranormal*, 59.
6 Sagan, *The Demon-Haunted World*, 302.
7 Swanson, *The Synchronized Universe: New Science of the Paranormal*, 59.
8 Ibid., 61.
9 Kaku, *Physics of the Impossible*, 92.
10 Ibid., 93.
11 Rosenblum and Kuttner, *Quantum Enigma*, 255.
12 Swanson, *The Synchronized Universe: New Science of the Paranormal*, 59.
13 Ibid., 60.
14 Herr, *Consciousness*, 2. For more information on studies related to this phenomenon, see Bosch, Steinkamp & Boller, *Examining psychokinesis: The interaction of human intention with random number generators—a meta-analysis*. Radin, Nelson, Dobyns, & Houtkooper, J. *Re-examining psychokinesis: Commentary on the Bösch, Steinkamp and Boller meta-analysis*. Nelson, Radin, Shoup & Bancel, *Correlations of continuous random data with major world events*. Also see global-mind.org/results.html.
15 Pearsall, *The Heart's Code*, 48.
16 *The Global Consciousness Project Meaningful Correlations in Random Data*, http://noosphere.princeton.edu/.
17 Radin, *Entangled Minds*, 197-207 and http://noetic.org/research/projects/mindatlarge.
18 McTaggart, *The Intention Experiment*, 181.
19 Nelson, 'Detecting Mass Consciousness: Effects of Globally Shared Attention and Emotion' in *How Consciousness Became the Universe*, Chopra et al., 107.
20 Allen, *How Uri Geller convinced the CIA he was a 'psychic warrior'*, http://www.telegraph.co.uk/news/2017/01/18/uri-geller-convinced-cia-psychic-warrior/.
21 Swanson, *The Synchronized Universe: New Science of the Paranormal*, 56.
22 Jacobsen, *Phenomena*, 403.
23 As cited in Swanson, *The Synchronized Universe: New Science of the Paranormal*, 53
24 Ibid., 58.
25 Targ, *The Reality of ESP*, 164.
26 Ibid., 164-166.
27 Lilou Mace. "Psychoenergetics – William Tiller Ph.D." YouTube video, 43:41. Posted October 15, 2012. https://www.youtube.com/watch?v=pI7jO1JuF-c&app=desktop.
28 Ibid.
29 Herr, *Consciousness*, 101.
30 Tiller, *A Brief Introduction to Intention-Host Device Research*. Also see https://www.youtube.com/watch?v=pI7jO1JuF-c.
31 Tiller, *Psychoenergetic Science: A Second Copernican-scale Revolution*, 14.
32 Targ, *The Reality of ESP*, 153.
33 Ibid., 170.

34 Diamond, 'Renaissance Man,' former UCI professor dies, http://www.latimes.com/tn-cpt-me-0712-joie-obit-20130711-story.html.
35 Jones, *An Extensive Laboratory Study of Pranic Healing Medical Imaging and Laboratory Methods*. Also see Swanson, *Life Force, the Scientific Basis: Volume 2 of the Synchronized Universe*, 21-25.
36 Swanson, *Life Force, the Scientific Basis: Volume 2 of the Synchronized Universe*, 24.
37 Ibid., 33.
38 Ibid., 32. Also see http://www.naturalhealingcenter.com/creative/jixingli.htm.

Chapter 9

1 Van Lommel, *Consciousness Beyond Life*, 162.
2 Laszlo, *What is Reality?*, 31.
3 Open Sciences website, "Gary Schwartz" (see video: *Is Consciousness More Than the Brain?*), http://opensciences.org/gary-schwartz.
4 Van Lommel, *Consciousness beyond Life*, 157.
5 ExpandedBooks. "Allan J. Hamilton—The Scalpel and the Soul." YouTube video, 3:47. Posted March 14, 2008. https://www.youtube.com/watch?v=rlfYnuNR3lI.
6 International Association for Near-Death Studies, Inc, *Key Facts about Near-Death Experiences*, http://iands.org/ndes/about-ndes/key-nde-facts21.html?showall=&limitstart=.
7 Van Lommel, *Consciousness Beyond Life*, 111.
8 Schwartz, 'The New Map in the Study of Consciousness' in *What is Reality?*, Laszlo, 135.
9 Near-Death Experience Research Foundation, http://www.nderf.org/index.htm.
10 Holden, Greyson, and James, 'The Field of Near-Death Studies: Past, Present, and Future' in *The Handbook of Near-Death Experiences*, ed. Holden et al., 2.
11 Ibid., 3.
12 Ibid., 2.
13 Swanson, *The Synchronized Universe: New Science of the Paranormal*, 189.
14 Moody, *Life After Life*, 7-8.
15 E.W. Kelly et al., 'Unusual Experiences Near Death and Related Phenomena' in Irreducible Mind, Kelly & Kelly et al., 373.
16 Moorjani, *The Day That My Life Changed: February 2, 2006*, http://anitamoorjani.com/about-anita/near-death-experience-description/.
17 Ibid.
18 Van Lommel, *Consciousness Beyond Life*, 59.
19 Ibid., 62.
20 IANDSvideos. "Jan Holden—NDE as Passage into Spontaneous Mediumship Experiences" YouTube video, 57:25. Posted March 4, 2016. https://www.youtube.com/watch?v=mxurOz0GU_A&app=desktop.
21 Ring, *Lessons from the Light: What We Can Learn from the Near-Death Experience*, 129.
22 Adapted from *Consciousness beyond Life* by Pim van Lommel, 17-41, and *Evidence of the Afterlife* by Jeffrey Long, 7-17.
23 Van Lommel, *Consciousness Beyond Life*, 18.
24 Nelson, 'Neuroscience Perspectives in Near-Death Experiences' in *The Science of Near-Death Experiences*, ed. Hagan, 114.
25 Cicoria and Cicoria, 'My Near-Death Experience: A Call from God' in *The Science of Near-Death Experiences*, ed. Hagan, 57.

26 Moorjani, *The Day That My Life Changed: February 2, 2006,* http://anitamoorjani.com/about-anita/near-death-experience-description/.
27 Swanson, *The Synchronized Universe: New Science of the Paranormal*, 190.
28 Long, *Evidence of the Afterlife*, 8.
29 Ibid., 60.
30 Cicoria and Cicoria, 'My Near-Death Experience: A Call from God' in *The Science of Near-Death Experiences*, ed. Hagan, 56.
31 Long, *Evidence of the Afterlife*, 14.
32 Ibid., 15.
33 Ibid., 9.
34 Van Lommel, *Consciousness Beyond Life*, 33.
35 Ibid., 32.
36 Cicoria and Cicoria, 'My Near-Death Experience: A Call from God' in *The Science of Near-Death Experiences*, ed. Hagan, 56-57.
37 Long, *Evidence of the Afterlife*, 14.
38 Van Lommel, *Consciousness Beyond Life*, 35.
39 Ibid., 36.
40 Long, *Evidence of the Afterlife*, 15.
41 Van Lommel, *Consciousness Beyond Life*, 38.
42 Cicoria and Cicoria, 'My Near-Death Experience: A Call from God' in *The Science of Near-Death Experiences*, ed. Hagan, 57.
43 Long, 'Near-Death Experiences: Evidence for their Reality' in *The Science of Near-Death Experiences*, ed. Hagan, 65.
44 E.W. Kelly et al., 'Unusual Experiences Near Death and Related Phenomena' in Irreducible Mind, Kelly & Kelly et al., 416.
45 Long, 'Near-Death Experiences: Evidence for their Reality' in *The Science of Near-Death Experiences*, ed. Hagan, 69.
46 Greyson, 'An Overview of Near-Death Experiences' in *The Science of Near-Death Experiences*, ed. Hagan, 21.
47 Long, 'Near-Death Experiences: Evidence for their Reality' in *The Science of Near-Death Experiences*, ed. Hagan, 73.
48 Van Lommel, *Consciousness Beyond Life*, 71.
49 As cited in Long, 'Near-Death Experiences: Evidence for their Reality' in *The Science of Near-Death Experiences*, ed. Hagan, 73.
50 Barušs and Mossbridge, *Transcendent Mind*, 110-111.
51 Greyson, 'An Overview of Near-Death Experiences' in *The Science of Near-Death Experiences*, ed. Hagan, 21.
52 Van Lommel, *Consciousness Beyond Life*, 148.
53 Barušs and Mossbridge, *Transcendent Mind*, 107.
54 E.W. Kelly et al., 'Unusual Experiences Near Death and Related Phenomena' in Irreducible Mind, Kelly & Kelly et al., 379-380.
55 Ibid., 380.
56 Harris, *Waking Up*, 182.
57 E.W. Kelly et al., 'Unusual Experiences Near Death and Related Phenomena' in Irreducible Mind, Kelly & Kelly et al., 381.
58 Ibid.
59 Harris, *Waking Up*, 180.
60 Alexander, *Proof of Heaven*, 186.
61 Ibid.

62 Ibid., 142.
63 Nelson, 'Neuroscience Perspectives in Near-Death Experiences' in *The Science of Near-Death Experiences*, ed. Hagan, 119.
64 Alexander, 'Near-Death Experiences and the Emerging Scientific View of Consciousness' in *The Science of Near-Death Experiences*, ed. Hagan, 127.
65 Van Lommel, *Near-Death Experience, Consciousness, and the Brain.*
66 Long, 'Near-Death Experiences: Evidence for their Reality' in *The Science of Near-Death Experiences*, ed. Hagan, 71.
67 Ibid.
68 Van Lommel, *Consciousness Beyond Life*, 20.
69 Long, *Evidence of the Afterlife*, 76.
70 Barušs and Mossbridge, *Transcendent Mind*, 116.
71 Moody, *Life After Life*, 96.
72 Ibid.
73 Moorjani, *The Day That My Life Changed: February 2, 2006*, http://anitamoorjani.com/about-anita/near-death-experience-description/.
74 International Association for Near-Death Studies, *Near-Death Experiences: Key Facts.*
75 Van Lommel, *Near-Death Experience, Consciousness, and the Brain.*
76 Van Lommel, *Consciousness Beyond Life*, 21-22.
77 Ibid., 166.
78 E.W. Kelly et al., 'Unusual Experiences Near Death and Related Phenomena' in Irreducible Mind, Kelly & Kelly et al., 418.
79 Van Lommel, *Consciousness Beyond Life*, 142.
80 Ibid., 166.
81 Van Lommel et al., *Near-death experience in survivors of cardiac arrest: a prospective study in the Netherlands*, http://www.thelancet.com/journals/lancet/article/PIIS0140673601071008/fulltext.
82 For a summary of these studies, see van Lommel, *Consciousness Beyond Life*, 156.
83 International Association for Near Death Studies, Inc, *AWARE study initial results are published*, https://iands.org/resources/media-resources/front-page-news/1060-aware-study-initial-results-are-published.html.
84 Ibid.
85 Harris, *Waking Up*, 173.
86 Ibid., 179.
87 Keim, *Consciousness After Death: Strange Tales From the Frontiers of Resuscitation Medicine*, https://www.wired.com/2013/04/consciousness-after-death/.
88 Grof, *Books of the Dead*, 31.
89 Ring and Cooper, *Mindsight*, 11.
90 Ibid., 13.
91 Ibid., 14.
92 Ibid., 15.
93 Ibid., 31.
94 Ibid., 32.
95 Ibid., 35.
96 Ibid., 84.
97 Ibid., 19.
98 Ibid., 41.

99 Ibid., 24.
100 Ibid., 82.
101 Weiss, *Same Soul, Many Bodies*, 9-10.
102 Van Lommel, *Consciousness Beyond Life*, 19.
103 Ibid., 23.
104 Ring and Cooper, *Mindsight*, 123.
105 Moody, *Life After Life*, 178.
106 Ibid., xiii.
107 Moody, *Glimpses of Eternity*, 11.
108 Ibid., 7.
109 Ibid., 13.
110 Ibid., 49.
111 Ibid., 50.
112 Ibid., 104.
113 Ibid., 87.
114 Ibid., 92.
115 Ibid., 71.
116 Ibid., 51.
117 Van Lommel, *Consciousness Beyond Life*, 8.
118 Ibid., 111.
119 E.W. Kelly et al., 'Unusual Experiences Near Death and Related Phenomena' in *Irreducible Mind*, Kelly & Kelly et al., 372.
120 The notion of "obfuscation" is one addressed by Bernardo Kastrup, e.g., see *Why Materialism Is Baloney* and *Brief Peeks Beyond*.

Chapter 10

1 Kastrup, *Why Materialism Is Baloney*, 184.
2 Open Sciences website, "Gary Schwartz" (see video: *Is Consciousness More Than the Brain?*), http://opensciences.org/gary-schwartz.
3 Braude, *Immortal Remains*, xiv.
4 Gauld, *Mediumship and Survival*, 261.
5 Myers, *Human Personality and Its Survival of Bodily Death*, 404.
6 Beischel, 'Research into Mental Mediumship' in *Surviving Death*, Kean, 173.
7 Braude, *Immortal Remains*, 306.
8 Gauld, *Mediumship and Survival*, 33.
9 Ibid., 34.
10 Ibid., 43.
11 Ibid., 42-43.
12 Myers et al., *A Record of Observations of Certain Phenomena of Trance*, 653.
13 Gauld, *Mediumship and Survival*, 45.
14 Ibid., 48.
15 Haraldsson and Stevenson, *A Communicator of the 'Drop In' Type in Iceland: The Case of Runolfur Runolfsson*.
16 Braude, *Immortal Remains*, 301.
17 Beischel, *Investigating Mediums*, 97.
18 Ibid., 88.
19 Beischel, 'Research into Mental Mediumship' in *Surviving Death*, Kean, 173.
20 Beischel et al., *Anomalous Information Reception by Research Mediums Under Blinded Conditions II: Replication and Extension*.

21 Beischel, 'Research into Mental Mediumship' in *Surviving Death*, Kean, 177.
22 Kean, *Surviving Death*, 187.
23 Ibid., 178.
24 Van Lommel, *Consciousness Beyond Life*, 294.
25 Davids, *An Atheist in Heaven*, 26.
26 Ibid.,126.
27 Ibid., Preface.
28 Ibid., 238.
29 Ibid., 242.
30 Ibid., 15.
31 Ibid., 17.
32 Guggenheim and Guggenheim, *Hello from Heaven!*, 244.
33 Pearson, *The stories of dying patients and doctors, will transform the way you think about your final days*, https://www.sott.net/article/279320-The-stories-of-dying-patients-and-doctors-will-transform-the-way-you-think-about-your-final-days.
34 Van Lommel, *Consciousness Beyond Life*, 292.
35 E.W. Kelly et al., 'Unusual Experiences Near Death and Related Phenomena' in *Irreducible Mind*, Kelly & Kelly et al., 408.

Chapter 11

1 Tucker, *Return to Life*, 165.
2 Kelly, 'Introduction: Science and Spirituality at a Crossroads' in Beyond Physicalism, Kelly et al., 9.
3 Stevenson, *Reincarnation and Biology: Volume I*, 1145.
4 Sagan, *The Demon-Haunted World*, 302.
5 Skeptiko, *The Dalai Lama is loved by millions, so why is this science professor demanding he step down? |270|*, https://skeptiko.com/270-lawrence-krauss-calls-for-dalai-lama-to-step-down/.
6 Stevenson, *Children Who Remember Previous Lives*, 13.
7 Dossey, *One Mind*, 90.
8 Ibid., 110.
9 As cited in Tucker, *Return to Life*, 13.
10 Stevenson, *Children Who Remember Previous Lives*, 110.
11 Stevenson, *Where Reincarnation and Biology Intersect*, 8.
12 Ibid., 9.
13 Stevenson, *Children Who Remember Previous Lives*, 109-110.
14 Ibid., 116.
15 Stevenson, *Where Reincarnation and Biology Intersect*, 7.
16 Stevenson, *Children who Remember Previous Lives*, 127.
17 Tucker, *Return to Life*, 63-87.
18 Ibid., 88-119.
19 Ibid., 94.
20 Ibid., 108.
21 Ibid., 112-113.
22 Stevenson, *Where Reincarnation and Biology Intersect*, 38.
23 Ibid., 41.
24 Ibid., 44.
25 Ibid., 43-44.

26 Stevenson, *Where Reincarnation and Biology Intersect*, 79.
27 Ibid., 79-80.
28 Ibid., 137.
29 Ibid., 138.
30 Ibid.
31 Ibid., 24.
32 Ibid., 25.
33 Ibid., 24.
34 Ibid., 24-27.
35 Stevenson, *Children Who Remember Previous Lives*, 12.

Chapter 12

1 Shermer, *Why Climate Skeptics Are Wrong*, https://www.scientificamerican.com/article/why-climate-skeptics-are-wrong/.
2 Mishlove, *The PK Man*, 5.
3 Sidgwick, *Address by the President at the first general meeting*, 12.
4 Eysenck, *Sense and Nonsense in Psychology.*
5 Utts, *An Assessment of Evidence for Psychic Functioning.*
6 Hyman, *Evaluation of Program on Anomalous Mental Phenomena*, http://www.ics.uci.edu/~jutts/hyman.html.
7 Tressoldi, *Extraordinary claims require extraordinary evidence: the case of nonlocalnonlocalnonlocal perception, a classical and Bayesian review of evidences.*
8 Kelly, 'Introduction: Science and Spirituality at a Crossroads' in *Beyond Physicalism*, Kelly et al., 4.
9 Utts, *An Assessment of Evidence for Psychic Functioning.*
10 Carey, A *Princeton Lab on ESP Plans to Close its Doors*, http://www.nytimes.com/2007/02/10/science/10princeton.html?mcubz=0.
11 Weiss, *Many Lives, Many Masters*, 128.
12 Ibid., 129.
13 Barušs and Mossbridge, *Transcendent Mind*, 25.
14 Carey, A *Princeton Lab on ESP Plans to Close its Doors*, http://www.nytimes.com/2007/02/10/science/10princeton.html?mcubz=0.
15 Jahn et al., *Correlations of Random Binary Sequences with Pre-Stated Operator Intention: A Review of a 12-Year Program.*
16 Josephson, *Coupled superconductors and beyond*, 261.
17 Crookes, *Researches in the Phenomena of Spiritualism.*
18 Oliver, *Marcello Truzzi, 67; Professor Studied the Far-Out From Witchcraft to Psychic Powers*, http://articles.latimes.com/2003/feb/11/local/me-truzzi11.
19 Sheldrake, *Wikipedia Under Threat*, https://www.sheldrake.org/about-rupert-sheldrake/blog/wikipedia-under-threat.
20 Radin, *The Conscious Universe*, 3.
21 Dabney, 'Maternal Impressions.' in *Cyclopedia of the Diseases of Children: Medical and Surgical*, ed. Keating, 191.
22 Beauregard, *Brain Wars*, 138.
23 Dan Jacob. "Neil deGrasse on consciousness." YouTube video, 1:14. Posted June 26, 2015. https://www.youtube.com/watch?v=QGekFhbyQLk.
24 Paulson, *Roger Penrose On Why Consciousness Does Not Compute*, http://nautil.us/issue/47/consciousness/roger-penrose-on-why-consciousness-does-not-compute.

25 Warman, *Stephen Hawking tells Google 'philosophy is dead'*, http://www.telegraph.co.uk/technology/google/8520033/Stephen-Hawking-tells-Google-philosophy-is-dead.html.
26 Skeptiko, *The Dalai Lama is loved by millions, so why is this science professor demanding he step down? |270|*, https://skeptiko.com/270-lawrence-krauss-calls-for-dalai-lama-to-step-down/.
27 Beischel et al., *Anomalous Information Reception by Research Mediums Under Blinded Conditions II: Replication and Extension.*
28 Dossey, *One Mind*, xxxviii.
29 Penrose and Hameroff, *Consciousness in the Universe: Neuroscience, Quantum Space-Time Geometry and Orch OR Theory.*
30 *Through the Wormhole*, "Is there life after death?" Season 2, episode 1. Science Channel, June 8, 2011.
31 Ball, *Roger Penrose maths genius; his ideas almost too out-there for some scientists*, http://www.afr.com/technology/roger-penrose-maths-genius-his-ideas-almost-too-outthere-for-some-scientists-20170220-gugubu.
32 Paulson, *Roger Penrose On Why Consciousness Does Not Compute*, http://nautil.us/issue/47/consciousness/roger-penrose-on-why-consciousness-does-not-compute.
33 Carroll, *Telekinesis and Quantum* Theory, http://www.preposterousuniverse.com/blog/2008/02/18/telekinesis-and-quantum-field-theory/.
34 Schwartz, *The Sacred Promise*, 267.
35 Radin, *The Conscious Universe*, 335.
36 As cited in Russell, *From Science to God*, 17.
37 Haisch, 'Reductionism and Consciousness' in *Mind before matter: Visions of a new science of consciousness*, Pfeiffer et al., 53.

Chapter 13

1 Radin, *Entangled Minds*, 291.
2 Targ, *The Reality of ESP*, 83.
3 Radin, *Supernormal*, 283.
4 Stevenson, *Children Who Remember Previous Lives*, 46.
5 Stott, *The Float Tank Cure*, 38.
6 Powell, *The ESP Enigma*, 129.
7 Urban, *Neuralink and the Brain's Magical Future*, https://waitbutwhy.com/2017/04/neuralink.html.
8 Stat, *Elon Musk launches Neuralink, a venture to merge the human brain with AI*, https://www.theverge.com/2017/3/27/15077864/elon-musk-neuralink-brain-computer-interface-ai-cyborgs.
9 Urban, *Neuralink and the Brain's Magical Future*, https://waitbutwhy.com/2017/04/neuralink.html.
10 Ibid.
11 Kurzweil, *Singularity Q&A*, http://www.kurzweilai.net/singularity-q-a.
12 Targ, *The Reality of ESP*, 262.
13 Sheldrake, *Dogs That Know When Their Owners Are Coming Home*, 273-286.
14 Ibid., 273-274.
15 Ibid., 270.
16 Ibid., 271.
17 Schwartz, *Opening to the Infinite*, 269.

18 Schnabel, *Remote Viewers*, 171.
19 Ibid.
20 Targ, *The Reality of ESP*, 262.
21 *Global Harmony Replication*, http://teilhard.global-mind.org/papers/pdf/global.harmony.html.
22 Herr, *Consciousness*, 52.
23 Currivan, *The Cosmic Hologram*, 8.
24 McDaniels, *Psychedelics reduce anxiety, depression in patients, study finds*, http://www.baltimoresun.com/health/bs-hs-psychedelics-cancer-20161201-story.html; also see http://www.collective-evolution.com/2016/12/02/study-single-session-of-ayahuasca-can-defeat-depression/.
25 Levy, *The Drug of Choice for the Age of Kale*, https://www.newyorker.com/magazine/2016/09/12/the-ayahuasca-boom-in-the-u-s.
26 Lipton, *The Biology of Belief*, 96.
27 Lipton and Bhaerman, *Spontaneous Evolution*, 28.
28 Church, *The Genie in Your Genes*, 48.
29 Ibid., 73.
30 Ibid., 75.
31 Moorjani, *The Day That My Life Changed: February 2, 2006*, http://anitamoorjani.com/about-anita/near-death-experience-description/.
32 Hawkins, *The Eye of the I*, 301.
33 Long, *Evidence of the Afterlife*, 9.
34 Ibid., 15.
35 Ibid., 8.
36 Swanson, *The Synchronized Universe: New Science of the Paranormal*, 19.
37 Hugenot, *The New Science of Consciousness Survival*, 73.
38 Elisabeth Kübler-Ross Foundation, *50 Quotes by Dr. Elisabeth Kübler-Ross*, http://www.ekrfoundation.org/quotes/.
39 Laszlo, *What is Reality?*, 263.
40 Van Lommel, *Consciousness Beyond Life*, 219.
41 Interview with Max Planck, The Observer.
42 Tucker, *Return to Life*, 191.
43 Ibid., 193.
44 Alexander, *Proof of Heaven*, 150.
45 Lanza and Berman, *Biocentrism*, 178.
46 Ibid.
47 Penrose and Hameroff, *Consciousness in the Universe: Neuroscience, Quantum Space-Time Geometry and Orch OR Theory*.
48 Conn Henry, *The mental Universe*.
49 Backster, *Primary Perception*, 141.
50 Weinberg, *Dreams of a Final Theory*.
51 Chopra et al., *How Consciousness Became the Universe*.
52 Gosling, *Science and the Indian Tradition: When Einstein Met Tagore*, 162.
53 Spira, *The Nature of Consciousness*, 125.
54 Ibid., 88.
55 For a fuller discussion of issues around the discussed form of panpsychism, see Dr. Bernardo Kastrup's *Why Materialism Is Baloney*, page 66 and Rupert Spira's The *Nature of Consciousness*, Chapter 3.
56 See Kastrup, *Brief Peeks Beyond*, 21-36.

57 Maharaj, *I am That*, 35.
58 Ibid., 9.
59 Hawkins, *I: Subjectivity and Reality*, 310.
60 Alexander, *Proof of Heaven*, 76.
61 Schrödinger, *What Is Life? with Mind and Matter*, 145.
62 Ibid., 165.
63 Moorjani, *Dying to Be Me*, 64.
64 Van Lommel, *Near-Death Experience, Consciousness, and the Brain.*
65 Radin, *Entangled Minds*, 3.
66 Haramein, 'The Physics of Oneness' in *What is Reality?*, Laszlo, 113.
67 Schrödinger, *What Is Life? with Mind and Matter*, 139.
68 Russell, *From Science to God*, 42.
69 Laszlo, *The Intelligence of the Cosmos*, 12.
70 Kastrup, *More Than Allegory*, 105.
71 Ibid., 106.
72 Ibid.
73 Ibid., 116-117.
74 Ibid., 99.
75 Ibid., 104.
76 Spira, *The Nature of Consciousness*, 156.
77 Conceptions of "now" and "here" have been addressed by many, but the discussion here has been inspired primarily by the work of Rupert Spira and Dr. Bernardo Kastrup. For example, if you enjoyed this thought exercise, you might enjoy the following talk from Rupert Spira, available at: https://www.youtube.com/watch?v=P00lv_bdNmo ("Rupert Spira Explaining the NOW Very Clearly" on Profound Talks, posted August 23, 2017). The exercise I provide for "now" and "here" was adapted from talks such as this.
78 Jahn and Dunne, *Consciousness and the Source of Reality*, 237.
79 Profound Talks. "Rupert Spira Explaining the NOW Very Clearly." YouTube video, 37:34. Posted August 23, 2017. https://www.youtube.com/watch?v=P00lv_bdNmo&feature=youtube.
80 Spira, *The Nature of Consciousness*, 160.
81 Ibid., 161.
82 Conn Henry, *The Mental Universe.*
83 Hawkins, *Discovery of the Presence of God*, 98.
84 Kastrup, *Why Materialism Is Baloney*, 198.
85 Ibid.
86 Hawkins, *The Eye of the I*, 142.
87 This introspection exercise was inspired by and adapted from the work of Rupert Spira. For example, the exercise I describe was heavily influenced by the following recording: Profound Talks. "Rupert Spira—The Highest Meditation (Beautiful Talk)." YouTube video, 53:49. Posted October 23, 2017. https://www.youtube.com/watch?v=ZxvlVOe1-6E&t=1830s.
88 Moorjani, *The Day That My Life Changed: February 2, 2006,* http://anitamoorjani.com/about-anita/near-death-experience-description/.
89 Kastrup, *Why Materialism Is Baloney*, 198.
90 Spira, *The Nature of Consciousness*, 165.
91 Schrödinger, *My View of the World*, 31-34.
92 Spira, *The Nature of Consciousness*, 140.

93 Hawkins, *Power vs. Force*, 384-385.
94 Ibid., 385-386.
95 Ibid., 385.
96 Hawkins, *I: Reality and Subjectivity*, 261.
97 Coppel, *The Awakening of Eckhart* Tolle, https://www.eckharttolle.com/article/Spiritual-Awakening-Of-Eckhart-Tolle.
98 Ubuntu Chan, *Superwellness*, 38.
99 Greenwell, *The Kundalini Guide*, 7-8.
100 Ibid., 13-14.
101 Kushlev et al., *Higher Income Is Associated With Less Daily Sadness but not More Daily Happiness*, http://journals.sagepub.com/doi/pdf/10.1177/1948550614568161.
102 Spira, *The Nature of Consciousness*, 193.
103 Coppel, *The Awakening of Eckhart* Tolle, https://www.eckharttolle.com/article/Spiritual-Awakening-Of-Eckhart-Tolle.
104 As cited in Price and Barrell, *Inner Experience and Neuroscience: Merging Both Perspectives*, 269.
105 Kastrup, *Brief Peeks Beyond*, 148.
106 Spira, *The Nature of Consciousness*, 150.
107 Ibid., 5.
108 Dossey, *One Mind*, xxi-xxii.

Bibliography

Alexander, Eben. "Near-Death Experiences and the Emerging Scientific View of Consciousness." In *The Science of Near-Death Experiences*, edited by John Hagan. Columbia: University of Missouri Press, 2017.

Alexander, Eben. "Near-Death Experiences: The Mind-Body Debate and the Nature of Reality." In *The Science of Near-Death Experiences*, edited by John Hagan. Columbia: University of Missouri Press, 2017.

Alexander, Eben. *Proof of Heaven: A Neurosurgeon's Journey into the Afterlife*. New York: Simon & Schuster, 2012.

Allen, Nick. "How Uri Geller Convinced the CIA He Was a 'Psychic Warrior.'" *The Telegraph*, January 18, 2017. http://www.telegraph.co.uk/news/2017/01/18/uri-geller-convinced-cia-psychic-warrior/.

Ananthaswamy, Anil. "A Classic Quantum Test Could Reveal the Limits of the Human Mind." *New Scientist*, May 19, 2017. https://www.newscientist.com/article/2131874-a-classic-quantum-test-could-reveal-the-limits-of-the-human-mind/.

Backster, Cleve. *Primary Perception: Biocommunication with Plants, Living Foods, and Human Cells*. Anza, CA: White Rose Millennium, 2003.

Ball, Philip. "Roger Penrose Maths Genius; His Ideas Almost Too Out-There For Some Scientists." *Financial Review*, February 28, 2017. http://www.afr.com/technology/roger-penrose-maths-genius-his-ideas-almost-too-outthere-for-some-scientists-20170220-gugubu.

Barušs, Imants, and Julia Mossbridge. *Transcendent Mind: Rethinking the Science of Consciousness*. Washington, DC: American Psychological Association, 2017.

Beauregard, Mario. *Brain Wars: The Scientific Battle Over the Existence of the Mind and the Proof That Will Change the Way We Live Our Lives*. Toronto, Canada: HarperCollins, 2013.

Beischel, Julie. *Investigating Mediums: A Windbridge Institute Collection*. Tuscon, AZ: Windbridge Institute, 2015.

Beischel, Julie. "Research into Mental Mediumship." In *Surviving Death: A Journalist Investigates Evidence for an Afterlife*, edited by Leslie Kean. New York: Crown Archetype, 2017.

Beischel, Julie, et al. "Anomalous Information Reception by Research Mediums Under Blinded Conditions II: Replication and Extension." *EXPLORE: The Journal of Science and Healing* 11, no. 2 (2015): 136–42.

Bell, J. S. "On the Einstein Podolsky Rosen Paradox." *Physics* 1, no. 3 (1964): 195–200.

Bem, Daryl. "Feeling the Future: Experimental Evidence for Anomalous Retroactive Influences on Cognition and Affect." *Journal of Personality and Social Psychology* 100, no. 3 (2011): 407–25.

Bem, Daryl, and Charles Honorton. "Does Psi Exist? Replicable Evidence for an Anomalous Process of Information Transfer." *Psychological Bulletin* 115 (1994): 4–18.

Bem, Daryl, et al. "Feeling the Future: A Meta-Analysis of 90 Experiments on the Anomalous Anticipation of Random Future Events." *F1000Research* 4 (2015): 1188. doi: 10.12688/f1000research.7177.1) https://f1000research.com/articles/4-1188/v1.

Bohm, D. J., and B. J. Hiley. "On the Intuitive Understanding of Nonlocality As Implied by Quantum Theory." *Foundations of Physics* 5, no. 1 (1975): 93–109.

Bosch, Holger, Fiona Steinkamp, and Emil Boller. "Examining Psychokinesis: The Interaction of Human Intention with Random Number Generators; A Meta-Analysis." *Psychological Bulletin* 132, no. 4 (2006): 497–523.

Braude, Stephen. *Immortal Remains: The Evidence for Life After Death*. Lanham, MD: Rowman & Littlefield, 2003.

Broderick, Damien, and Ben Goertzel, eds. *Evidence for Psi: Thirteen Empirical Research Reports*. Jefferson, NC: McFarland, 2015.

Brooks, Michael. "Beyond the Safe Zone of Science." *EdgeScience* 23 (2015).

Buchheit, Carl, and Ellie Nower Schamber. *Transformational NLP: A New Psychology*. Ashland, OR: White Cloud, 2017.

Burt, Cyril. *ESP and Psychology*. Edited by Anita Gregory. London: Weidenfeld and Nicolson, 1975.

Capra, Fritjof. *The Tao of Physics: An Exploration of the Parallels between Modern Physics and Eastern Mysticism*. Boston: Shambhala, 1975.

Cardeña, E. "The Experimental Evidence for Parapsychological Phenomena: A Review." *American Psychologist* (May 24, 2018). Advance online publication. http://dx.doi.org/10.1037/amp0000236.

Carey, Benedict. "Journal's Paper on ESP Expected to Prompt Outrage." *New York Times*, January 5, 2011. http://www.nytimes.com/2011/01/06/science/06esp.html.

Carey, Benedict. "A Princeton Lab on ESP Plans to Close Its Doors." *New York Times*, February 10, 2007. http://www.nytimes.com/2007/02/10/science/10princeton.html?mcubz=0.

Carhart-Harris, R. L., et al. "Neural Correlates of the Psychedelic State as Determined by fMRI Studies with Psilocybin." *Proceedings of the National Academy of Sciences USA* 109, no. 6 (2012): 2138–43. https://www.ncbi.nlm.nih.gov/pubmed/22308440

Carr, Bernard. "Hyperspatial Models of Matter and Mind." In *Beyond Physicalism: Toward Reconciliation of Science and Spirituality*, edited by Edward Kelly, Adam Crabtree, and Paul Marshall. Lanham, MD: Rowman & Littlefield, 2015.

Carroll, Sean. *The Big Picture: On the Origins of Life, Meaning, and the Universe Itself.* New York: Dutton, 2017.

Carroll, Sean. "Telekinesis and Quantum Field Theory." Sean Carroll website, February 18, 2008. http://www.preposterousuniverse.com/blog/2008/02/18/telekinesis-and-quantum-field-theory/.

Chan, Edith Ubuntu. *SuperWellness: Become Your Own Best Healer*. San Francisco, CA: School of Dan Tian Wellness, 2017.

Chopra, Deepak, et al. *How Consciousness Became the Universe: Quantum Physics, Cosmology, Evolution, Neuroscience, Parallel Universes.* Cambridge, MA: Cosmology Science, 2015.

Church, Dawson. *Genie in Your Genes: Epigenetic Medicine and the New Biology of Intention*. Santa Rosa, CA: Energy Psychology, 2014.

Cicoria, Tony, and Jordan Cicoria. "My Near-Death Experience: A Call From God." In *The Science of Near-Death Experiences*, edited by John Hagan. Columbia: University of Missouri Press, 2017.

Closer to Truth. "Lawrence Krauss: Does ESP Make Sense?" YouTube Video, July 17, 2017. https://youtu.be/5NweHLQmbZE.

"Collective Consciousness." IONS: Institute of Noetic Sciences website, n.d. May 24, 2018, http://noetic.org/research/projects/mindatlarge.

Coppel, Paula. "The Awakening of Eckhart Tolle." Eckhart Teachings website, n.d. https://www.eckharttolle.com/article/Spiritual-Awakening-Of-Eckhart-Tolle.

Costandi, Mo. "Psychedelic Chemical Subdues Brain Activity: Magic Mushrooms' Active Ingredient Constrains Control Centres." *Nature*, January 23, 2012. https://www.nature.com/news/psychedelic-chemical-subdues-brain-activity-1.9878.

Coyne, Jerry. "Science is Being Bashed by Academics Who Should Know Better." *New Republic*, April 3, 2014. https://newrepublic.com/article/117244/jeffrey-kripals-anti-materialist-argument-promotes-esp.

Crookes, William. *Researches in the Phenomena of Spiritualism*. Cambridge, UK: Cambridge University Press, 2012. First published 1874.

Currivan, Jude. *The Cosmic Hologram: In-Formation at the Center of Creation*. Rochester, VT: Inner Traditions, 2017.

Dabney, William. "Maternal Impressions." In *Cyclopedia of the Diseases of Children: Medical and Surgical*, edited by John Keating. Philadelphia, PA: J. B. Lippincott, 1890.

"The Dalai Lama is Loved By Millions, So Why Is This Science Professor Demanding He Step Down?" Skeptiko website, n.d. https://skeptiko.com/270-lawrence-krauss-calls-for-dalai-lama-to-step-down/.

Dalio, Ray. *Principles*. New York: Simon & Schuster, 2017.

Dan Jacob. "Neil deGrasse On Consciousness." YouTube Video, June 26, 2015. https://www.youtube.com/watch?v=QGekFhbyQLk.

Davids, Paul, Gary Schwartz, and John Allison. *An Atheist in Heaven: The Ultimate Evidence for Life After Death?* Reno, NV: Yellow Hat, 2016.

Diamond, Barbara. "'Renaissance Man,' Former UCI Professor Dies." *Los Angeles Times*, July 11, 2013. http://www.latimes.com/tn-cpt-me-0712-joie-obit-20130711-story.html.

Dossey, Larry. *One Mind: How Our Individual Mind Is Part of a Greater Consciousness and Why It Matters*. Carlsbad, CA: Hay House, 2013.

Dossey, Larry. *The Power of Premonitions: How Knowing the Future Can Shape Our Lives*. London: Hay House, 2009.

Dossey, Larry. "The Power of Premonitions: How Knowing the Future Can Shape Our Lives [interview]." n.d. http://www.dosseydossey.com/larry/Interview_Questions-Premonitions.pdf.

Dossey, Larry. "Unbroken Wholeness: The Emerging View of Human Interconnection." Reality Sandwich website, n.d. http://realitysandwich.com/170309/human_interconnection/.

Einstein, Albert, Hedwig Born, and Max Born. *The Born-Einstein Letters: Friendship, Politics and Physics in Uncertain Times*. Translated by Irene Born; note on the new edition by Gustav Born; new preface by Diana Buchwald and Kip Thorne; foreword by Bertrand Russell; introduction by Werner Heisenberg. Basingstoke, UK: Macmillan, 2005.

Einstein, Albert, Boris Podolsky, and Nathan Rosen. "Can Quantum-Mechanical Description of Physical Reality Be Considered Complete?" *Physical Review* 47, no. 777 (1935).

ExpandedBooks. "Allan J. Hamilton: The Scalpel and the Soul." YouTube Video, March 14, 2008. https://www.youtube.com/watch?v=rlfYnuNR3lI.

"Experiment Confirms Quantum Theory Weirdness." Phys.org. May 27, 2015. https://phys.org/news/2015-05-quantum-theory-weirdness.html.

"Experiments: Uri Geller at SRI, August 4–11, 1973." Unpublished CIA report, approved for release March 28, 2003.

"Ex-Stargate Head, Ed May, Unyielding Re Materialism, Slams Dean Radin." Skeptiko website, interview, n.d. http://skeptiko.com/ed-may-unyielding-re-materialism-341/.

Eysenck, H. J. *Sense and Nonsense in Psychology*. Middlesex, UK: Penguin, 1957.

Feynman, Richard. *The Character of Physical Law*. New York: Modern Library, 1994.

Feynman, Richard. "1. Quantum Behavior." Feynman Lectures Website, n.d. http://www.feynmanlectures.caltech.edu/III_01.html#Ch1-S7.

"Formal Results: Testing the GCP Hypothesis." Global Consciousness Project website, n.d. global-mind.org/results.html.

Friedman, Lauren. "A Neurosurgeon Calls This Basic Fact About The Brain 'Too Strange To Understand.'" *Business Insider*, June 24, 2016. http://www.businessinsider.com/the-strangest-thing-about-the-brain-2016-6.

"Ganzfeld experiment." Wikipedia, last updated August 13, 2017. https://en.wikipedia.org/wiki/Ganzfeld_experiment.

Gauld, Alan. *Mediumship and Survival: A Century of Investigations*. London: Paladin, 1983.

Gleick, James. *Chaos*. New York: Open Road Media, 1987.

"The Global Consciousness Project Meaningful Correlations in Random Data." Global Consciousness Project website, n.d. http://noosphere.princeton.edu/.

"Global Harmony Replication." Global Consciousness Project website, n.d. http://teilhard.global-mind.org/papers/pdf/global.harmony.html.

Gosling, David. *Science and the Indian Tradition: When Einstein Met Tagore*. London: Routledge, 2007.

Greene, Brian. *The Fabric of the Cosmos: Space, Time, and the Texture of Reality*. New York: A.A. Knopf, 2004.

Greene, Brian. *The Hidden Reality: Parallel Universes and the Deep Laws of the Cosmos*. New York: Vintage Books, 2013.

Greenwell, Bonnie. *The Kundalini Guide: A Companion for the Inward Journey*. Ashland, OR: Shakti River, 2014.

Greyson, Bruce. "An Overview of Near-Death Experiences." In *The Science of Near-Death Experiences*, edited by John Hagan. Columbia: University of Missouri Press, 2017.

Grof, Stanislav. *Books of the Dead*. New York: W. W. Norton, 2013.

Guerrer, Gabriel. "Consciousness-Related Interactions in a Double-Slit Optical System." *Open Science Framework* (March 9, 2018). doi:10.17605/OSF.IO/QDKVX.

Guggenheim, Bill, and Judy Guggenheim. *Hello from Heaven!* London: Watkins, 1995.

Hafele, J. C., and R. E. Keating. "Around-the-World Atomic Clocks: Predicted Relativistic Time Gains." *Science* 177 (1972): 166–68.

Hagan, John, ed. *The Science of Near-Death Experiences*. Columbia: University of Missouri Press, 2017.

Haisch, Bernard. "Reductionism and Consciousness." In *Mind Before Matter: Visions of a New Science of Consciousness*, edited by Trish Pfeiffer and John Mack. Winchester, UK: O Books, 2007.

Hameroff, Stuart, and Roger Penrose. "Consciousness in the Universe: A Review of the 'Orch OR' Theory." *Physics of Life Reviews* 11, no. 1 (2014): 39–78.

Haraldsson, Erlendur, and Ian Stevenson. "A Communicator of the 'Drop In' Type In Iceland: The Case of Runolfur Runolfsson." *Journal of the American Society for Psychical Research* 69 (1975): 33–59.

Haramein, Nassim. "The Physics of Oneness." In *What Is Reality: The New Map of Cosmos and Consciousness* by Ervin Laszlo. New York: SelectBooks, 2016.

Harari, Yuval Noah. *Homo Deus: A Brief History of Tomorrow*. London: Harvill Secker, 2017.

"The Hard Problem of Consciousness." *Internet Encyclopedia of Philosophy*. http://www.iep.utm.edu/hard-con/.

Harris, Sam. *Waking Up: A Guide to Spirituality Without Religion*. New York: Simon & Schuster, 2015.

Hawking, Stephen, and Leonard Mlodinow. *The Grand Design*. New York: Bantam, 2010.

Hawkins, David. *Discovery of the Presence of God: Devotional Nonduality*. West Sedona, AZ: Veritas, 2007.

Hawkins, David. *The Eye of the I: From Which Nothing Is Hidden*. Carlsbad, CA: Hay House 2001.

Hawkins, David. *I: Reality and Subjectivity*. Alexandria, Australia: Hay House, 2003.

Hawkins, David. *Power vs. Force: The Hidden Determinants of Human Behavior*. Carlsbad, CA: Hay House, 1995.

Heisenberg, Werner. *Physics and Philosophy: The Revolution in Modern Science*. New York: HarperPerennial, 1958.

Henry, Richard Conn. "The Mental Universe." *Nature* 436 (July 2005): 29.

Herr, Eva. *Consciousness: Bridging the Gap Between Conventional Science and the New Super Science of Quantum Mechanics*. Faber, VA: Rainbow Ridge, 2012.

Hofstadter, Douglas. "A Cutoff for Craziness." *New York Times*, January 7, 2011. https://www.nytimes.com/roomfordebate/2011/01/06/the-esp-study-when-science-goes-psychic/a-cutoff-for-craziness.

Holden, Janice Miner, Bruce Greyson, and Debbie James, eds. *The Handbook of Near-Death Experiences: Thirty Years of Investigation*. Santa Barbara, CA: Praeger, 2009.

Honorton, Charles, and Diane Ferrari. "'Future Telling'': A Meta-Analysis of Forced-Choice Precognition Experiments, 1935–1987." *Journal of Parapsychology* 53 (1989): 281–308.

Hugenot, Alan Ross. *The New Science of Consciousness Survival and the Meta-paradigm Shift to a Conscious Universe*. Indianapolis, IN: Dog Ear, 2016.

Hyman, Ray. "Evaluation of Program on Anomalous Mental Phenomena." Working paper, University of Oregon, Eugene, 1995. http://www.ics.uci.edu/~jutts/hyman.html.

IANDSvideos. "Jan Holden: NDE as Passage into Spontaneous Mediumship Experiences." YouTube Video, March 4, 2016. https://www.youtube.com/watch?v=mxurOz0GU_A&app=desktop.

InfinicityFilm. "Rupert Sheldrake: Dogs Who Know When Their Owners Are Coming Home." YouTube video, April 22, 2012. https://www.youtube.com/watch?v=9QsPWitQovM.

Institute of Noetic Sciences: IONS. "New Experiments Show Consciousness Affects Matter: Dean Radin Ph.D." YouTube Video, June 7, 2016. https://www.youtube.com/watch?v=nRSBaq3vAeY.

International Association for Near Death Studies. "AWARE Study Initial Results Are Published!" IANDS webpage, July 18, 2017. https://iands.org/resources/media-resources/front-page-news/1060-aware-study-initial-results-are-published.html.

International Association for Near Death Studies. "Key Facts about Near-Death Experiences." IANDS webpage, August 29, 2017. http://iands.org/ndes/about-ndes/key-nde-facts21.html?showall=&limitstart=.

"Interview with Max Planck." *The Observer*, January 25, 1931.

"Interview with Ray Hyman." *Austin American-Statesman*, July 14, 2002. As cited in Stephan A. Schwartz, *Opening to the Infinite: The Art and Science of Nonlocal Awareness* (Buda, TX: Nemoseen Media, 2007).

"Islands of Genius with Darold A. Treffert, M.D." Provocative Enlightenment podcast, August 10, 2017. http://provocativeenlightenment.com/wp/2017-0810-islands-of-genius-with-darold-a-treffert-m-d/.

Jacobsen, Annie. *Phenomena: The Secret History of the U.S. Government's Investigations into Extrasensory Perception and Psychokinesis*. New York: Black Bay, 2017.

Jahn, Robert, and Brenda Dunne. *Consciousness and the Source of Reality: The PEAR Odyssey*. Princeton, NJ: ICRL, 2011.

Jahn, Robert, and Brenda Dunne. *Margins of Reality: The Role of Consciousness in the Physical World*. Princeton, NJ: ICRL Press, 1981.

Jahn, Robert, et al. "Correlations of Random Binary Sequences with Pre-Stated Operator Intention: A Review of a 12-Year Program." *Explore* 3, no. 3 (2007): 244–53, 341–43. https://www.explorejournal.com/article/S1550-8307(07)00062-6/fulltext.

Jauch, J. M. "The Problem of Measurement in Quantum Mechanics." *Helvetica Physica Acta* 37 (1964): 293–316.

Jeans, James. *The Mysterious Universe*. New York: Macmillan, 1930.

Jones, Joie. "An Extensive Laboratory Study of Pranic Healing Medical Imaging and Laboratory Methods." Paper presented at the World Pranic Healers' Convention, Mumbai, India, May 12–14, 2006.

Josephson, Brian. "Coupled Superconductors and Beyond." *Low Temperature Physics* 38 (2012): 260–62. https://aip.scitation.org/doi/10.1063/1.3697974.

Josephson, Brian. "Pathological Disbelief." Lecture given at the Nobel Laureates' meeting Lindau, June 30, 2004, edited version of presentation (revised Aug. 20, 2004).

"Julia Mossbridge." IONS: Institute of Noetic Sciences website, n.d. http://noetic.org/directory/person/julia-mossbridge.

Kaku, Michio. *The Future of the Mind: The Scientific Quest to Understand, Enhance, and Empower the Mind*. New York: Random House, 2014.

Kaku, Michio. *Physics of the Impossible: A Scientific Tour Beyond Science Fiction, Fantasy, and Magic*. London: Allen Lane, 2008.

Kastrup, Bernardo. *Brief Peeks Beyond: Critical Essays on Metaphysics, Neuroscience, Free Will, Skepticism, and Culture*. Winchester, UK: Iff, 2015.

Kastrup, Bernardo. *More Than Allegory: On Religious Myth, Truth and Belief*. Winchester, UK: John Hunt 2016.

Kastrup, Bernardo. "Transcending the Brain: At Least Some Cases of Physical Damage Are Associated with Enriched Consciousness or Cognitive Skill." *Scientific American* blog, March 29, 2017. https://blogs.scientificamerican.com/guest-blog/transcending-the-brain/.

Kastrup, Bernardo. *Why Materialism Is Baloney: How True Skeptics Know There Is No Death and Fathom Answers to Life, the Universe and Everything*. Winchester, UK: Iff, 2014.

Kean, Leslie, ed. *Surviving Death: A Journalist Investigates Evidence for an Afterlife*. New York: Crown Archetype, 2017.

Keim, Brandon. "Consciousness After Death: Strange Tales From the Frontiers of Resuscitation Medicine." *Wired*, April 24, 2013. https://www.wired.com/2013/04/consciousness-after-death/.

Kelly, Edward. "Empirical Challenges to Theory Construction." In *Beyond Physicalism: Toward Reconciliation of Science and Spirituality*, edited by Edward Kelly, Adam Crabtree, and Paul Marshall. Lanham, MD: Rowman & Littlefield, 2015.

Kelly, Edward. "Toward a Psychology for the 21st Century." In *Irreducible Mind: Toward a Psychology for the 21st Century*, edited by Edward Kelly et al. Lanham, MD: Rowman & Littlefield, 2010.

Kelly, Edward, and David Presti. "A Psychobiological Perspective on 'Transmission Models.'" In *Beyond Physicalism: Toward Reconciliation of Science and Spirituality*, edited by Edward Lanham, Adam Crabtree, and Paul Marshall. Lanham, MD: Rowman & Littlefield, 2015.

Kelly, Edward, B. Greyson, and E. F. Kelly. "Unusual Experiences Near Death and Related Phenomena." In *Irreducible Mind: Toward a Psychology For the 21st Century*, edited by Edward Kelly et al. Lanham, MD: Rowman & Littlefield, 2010.

Kelly, Edward, et al. *Irreducible Mind: Toward a Psychology for the 21st Century*. Lanham, MD: Rowman & Littlefield, 2010.

Kelly, Edward, Adam Crabtree, and Paul Marshall, eds. *Beyond Physicalism: Toward Reconciliation of Science and Spirituality*. Lanham, MD: Rowman & Littlefield, 2015.

"Kim Peek" The Real Rain Man." Wisconsin Medical Society webpage, n.d. https://www.wisconsinmedicalsociety.org/professional/savant-syndrome/profiles-and-videos/profiles/kim-peek-the-real-rain-man/.

King, Barbara. "Interview: Richard Dawkins Celebrates Reason, Ridicules Faith." *NPR*, March 26, 2012. https://www.npr.org/sections/13.7/2012/03/26/149310560/atheist-firebrand-richard-dawkins-unrepentant-for-harsh-words-targeting-faith.

Knox, Sarah. *Science, God, and the Nature of Reality: Bias in Biomedical Research*. Boca Raton, FL: Brown Walker, 2010.

Kübler-Ross, Elisabeth. "Quotes: 50 Quotes by Dr. Elisabeth Kübler-Ross." Elisabeth Kübler-Ross Foundation webpage, n.d. http://www.ekrfoundation.org/quotes/.

Kurzweil, Ray. "Singularity Q&A." Kurzweil: Accelerating Intelligence website. December 9, 2011. http://www.kurzweilai.net/singularity-q-a.

Kushlev K., E. W. Dunn, and R. E. Lucas. "Higher Income Is Associated With Less Daily Sadness but Not More Daily Happiness." *Social Psychological and Personality Science* 6, no. 5 (2015): 483–89. http://journals.sagepub.com/doi/pdf/10.1177/1948550614568161

Lanza, Robert, and Bob Berman. *Beyond Biocentrism: Rethinking Time, Space, Consciousness, and the Illusion of Death*. Dallas, TX: BenBella, 2016.

Lanza, Robert, and Bob Berman. *Biocentrism: How Life and Consciousness Are the Keys to Understanding the True Nature of the Universe*. Dallas, TX: BenBella, 2009.

Laszlo, Ervin. *The Intelligence of the Cosmos: Why Are We Here? New Answers from the Frontiers of Science*. Rochester, VT: Inner Traditions, 2017.

Laszlo, Ervin. *What Is Reality: The New Map of Cosmos and Consciousness*. New York: SelectBooks, 2016.

Levy, Ariel. "The Drug of Choice for the Age of Kale: How Ayahuasca, an Ancient Amazonian Hallucinogenic Brew, Became the Latest Trend in Brooklyn and Silicon Valley." *New Yorker*, September 12, 2016. https://www.newyorker.com/magazine/2016/09/12/the-ayahuasca-boom-in-the-u-s.

Lilou Mace. "Psychoenergetics: William Tiller Ph.D." YouTube Video, October 15, 2012. https://www.youtube.com/watch?v=pI7jO1JuF-c.

Lipton, Bruce. *The Biology of Belief: Unleashing the Power of Consciousness, Matter, and Miracles*. Santa Rosa, CA: Mountain of Love/Elite Books, 2005.

Lipton, Bruce, and Steve Bhaerman. *Spontaneous Evolution: Our Positive Future (and a Way to Get There from Here)*. Carlsbad, CA: Hay House, 2012.

Long, Jeffrey. "Near-Death Experiences: Evidence for their Reality." In *The Science of Near-Death Experiences*, edited by John Hagan. Columbia: University of Missouri Press, 2017.

Long, Jeffrey, and Paul Perry. *Evidence of the Afterlife: The Science of Near-Death Experiences*. New York: HarperOne, 2010.

Lowery, George. "Study Showing That Humans Have Some Psychic Powers Caps Daryl Bem's Career." *Cornell Chronicle*, December 6, 2010. http://news.cornell.edu/stories/2010/12/study-looks-brains-ability-see-future.

MacIsaac, Tara. "Neuroscientist Discusses Precognition: Or 'Mental Time Travel.'" *Epoch Times*, November 22, 2017. https://www.theepochtimes.com/uplift/neuroscientist-discusses-precognition-or-mental-time-travel_2362702.html.

Maharaj, Nisargadatta, Maurice Frydman, and Sudhakar Dikshit. *I Am That: Talks with Sri Nisargadatta Maharaj*. Translated by Maureic Frydman. Durham, NC: Acorn Press, 1973.

"Manifesto for a Post-Materialist Science." OpenSciences.org website, n.d. http://opensciences.org/about/manifesto-for-a-post-materialist-science.

Markoff, John. "Sorry, Einstein: Quantum Study Suggests 'Spooky Action' Is Real." *New York Times*, October 21, 2015. https://www.nytimes.com/2015/10/22/science/quantum-theory-experiment-said-to-prove-spooky-interactions.html.

Mazza, Ed. "Virginia Man Wins The Lottery By Playing Numbers From A Dream: Some Dreams Really Do Come True." *Huffington Post*, February 6, 2018. https://www.huffingtonpost.com/entry/lottery-dream-numbers-jackpot_us_5a79402be4b00f94fe9456f0.

McDaniels, Andrea. "Psychedelics Reduce Anxiety, Depression in Patients, Study Finds." *Baltimore Sun*, December 1, 2016. http://www.baltimoresun.com/health/bs-hs-psychedelics-cancer-20161201-story.html.

McKie, Robin. "Royal Mail's Nobel Guru In Telepathy Row." *The Observer*, September 29, 2001. https://www.theguardian.com/uk/2001/sep/30/robinmckie.theobserver.

McTaggart, Lynne. *The Field: The Quest for the Secret Force of the Universe*. New York: HarperCollins, 2001.

McTaggart, Lynne. *The Intention Experiment: Using Your Thoughts to Change Your Life and the World*. New York: Free Press, 2007.

Mehra, Jagdish, ed. *The Physicist's Conception of Nature*. Boston: Reidel, 1973.

Mishlove, Jeffrey. *The PK Man: A True Story of Mind Over Matter*. Charlottesville, VA: Hampton Roads, 2000.

Moody, Raymond. *Glimpses of Eternity: An Investigation into Shared Death Experiences*. London: Rider, 2011.

Moody, Raymond. *Life After Life: The Investigation of a Phenomenon; Survival of Bodily Death*. Harrisburg, PA: Stackpole, 1975.

Moorjani, Anita. "The Day My Life Changed: February 2, 2006." Anita Moorjani webpage, n.d. http://anitamoorjani.com/about-anita/near-death-experience-description/.

Moorjani, Anita. *Dying to Be Me: My Journey from Cancer, to Near Death, to True Healing*. New Delhi, India: Hay House, 2015.

Mossbridge, Julia. "Physiological Activity That Seems to Anticipate Future Events." In *Evidence for Psi: Thirteen Empirical Research Reports*, edited by Damien Broderick and Ben Goertzel. Jefferson, NC: McFarland, 2015.

Mossbridge J., P. Tressoldi, and J. Utts. "Predictive Physiological Anticipation Preceding Seemingly Unpredictable Stimuli: A Meta-Analysis." *Frontiers in Psychology* 3 (2012): 390. https://www.frontiersin.org/articles/10.3389/fpsyg.2012.00390/full.

Mossbridge, Julia, et al. "Predicting the Unpredictable: Critical Analysis and Practical Implications of Predictive Anticipatory Activity." *Frontiers in Human Neuroscience* 8 (2014): 146. https://www.ncbi.nlm.nih.gov/pmc/articles/PMC3971164/.

Myers, F. W. H., and Richard Hodgson. *Human Personality and Its Survival of Bodily Death*. New York: Dover, 1910.

Myers. F. W. H., et al. "A Record of Observations of Certain Phenomena of Trance." *Proceedings of the Society for Psychical Research* 6 (1889–90): 436–659.

Nahm, M., et al. "Terminal Lucidity: A Review and a Case Collection." *Archives of Gerontology and Geriatrics* 55, no. 1 (2011): 138–42. doi:10.1016/j.archger.2011.06.031

NASA. "Tour of the Electromagnetic Spectrum: Introduction to the Electromagnetic Spectrum." NASA website, n.d. https://science.nasa.gov/ems/01_intro.

Near-Death Experience Research Foundation. http://www.nderf.org/index.htm.

Near-Death Experience Research Foundation. "Choose Your Language." n.d. http://www.nderf.org/NDERF/Languages/languages.htm.

Nelson, Kevin. "Neuroscience Perspectives on Near-Death Experiences." In *The Science of Near-Death Experiences*, edited by John Hagan. Columbia: University of Missouri Press, 2017.

Nelson, R. D., et al. "Correlations of Continuous Random Data With Major World Events." *Foundations of Physics Letters* 15 (2002): 537–50.

Noë, Alva. *Out of Our Heads: Why You Are Not Your Brain, and Other Lessons from the Biology of Consciousness*. New York: Hill & Wang, 2010.

"Notable Quotes on Quantum Physics." QuantumEnigma website, n.d. http://quantumenigma.com/nutshell/notable-quotes-on-quantum-physics/?phpMyAdmin=54029d98ba071eec0c69ff5c106b9539#einstein.

Oliver, Myrna. "Marcello Truzzi, 67; Professor Studied the Far-Out From Witchcraft to Psychic Powers." *Los Angeles Times*, February 11, 2003. http://articles.latimes.com/2003/feb/11/local/me-truzzi11.

"125th Anniversary Issue." *Science* magazine website. http://www.sciencemag.org/site/feature/misc/webfeat/125th/.

Parker, Adrian, and Göran Brusewitz. "A Compendium of the Evidence for Psi." *European Journal of Parapsychology* 18 (2003): 33–51.

Paulson, Steve. "Roger Penrose On Why Consciousness Does Not Compute: The Emperor of Physics Defends His Controversial Theory of Mind." *Nautilus*, May 4, 2017. http://nautil.us/issue/47/consciousness/roger-penrose-on-why-consciousness-does-not-compute.

Pearsall, Paul. *The Heart's Code: Tapping the Wisdom and Power of Our Heart Energy; The New Findings About Cellular Memories and Their Role in the Mind, Body, Spirit Connection*. New York: Broadway, 1999.

Pearson, Patricia. "At the Gates of Heaven: A New Book, Drawing On the Stories of Dying Patients and Doctors, Will Transform the Way You Think About Your Final Days." *Daily Mail*, May 16, 2014, http://www.dailymail.co.uk/news/article-2630927/At-gates-heaven-A-new-book-drawing-stories-dying-patients-doctors-transform-way-think-final-days.html.

Pearson, Patricia. "The Stories of Dying Patients and Doctors, Will Transform the Way You Think About Your Final Days." SOTT: Signs of the Times website, May 17, 2014. https://www.sott.net/article/279320-The-stories-of-dying-patients-and-doctors-will-transform-the-way-you-think-about-your-final-days.

Penman, Danny. "Could There Be Proof to the Theory That We're ALL Psychic?" *Daily Mail*, January 28, 2008. http://www.dailymail.co.uk/news/article-510762/Could-proof-theory-ALL-psychic.html.

Peoc'h, R. "Psychokinetic Action of Young Chicks on the Path of An Illuminated Source." *Journal of Scientific Exploration* 9 (1988): 223–29.

Pietsch, Paul. *Shufflebrain*. Boston: Houghton Mifflin, 1981.

Playfair, Guy Lyon. *Twin Telepathy*. London: Vega, 2012.

Powell, Diane Hennacy. *The ESP Enigma: The Scientific Case for Psychic Phenomena*. New York: Walker, 2010.

Powell, Diane, Paul Mills, and Deepak Chopra. "Non Local Consciousness in an Autistic Child." Paper presented at The Science of Consciousness meeting, San Diego, CA, June 5–10, 2017.

Price, Donald, and James Barrell. *Inner Experience and Neuroscience: Merging Both Perspectives*. Cambridge, MA: MIT Press, 2014.

Profound Talks. "Rupert Spira Explaining the NOW Very Clearly." YouTube Video, August 23, 2017. https://www.youtube.com/watch?v=P00lv_bdNmo.

Profound Talks. "Rupert Spira: The Highest Meditation (Beautiful Talk)." YouTube Video, October 23, 2017. https://www.youtube.com/watch?v=ZxvlVOe1-6E&t=1830s.

"Project Sun Streak." Unpublished CIA report, approved for release August 8, 2008.

Radin, Dean. "Biography." DeanRadin.com, n.d. http://www.deanradin.com/NewWeb/bio.html.

Radin, Dean. *The Conscious Universe: The Scientific Truth of Psychic Phenomena*. New York: HarperEdge, 1997.

Radin, Dean. *Entangled Minds Extrasensory Experiences in a Quantum Reality*. New York: Paraview, 2006.

Radin, Dean. *Real Magic: Ancient Wisdom, Modern Science, and a Guide to the Secret Power of the Universe*. New York: Harmony, 2018.

Radin, Dean. *Supernormal*. New York: Random House, 2013.

Radin, Dean. Twitter post. December 15, 2015. https://twitter.com/deanradin/status/676865032009814016.

Radin, Dean, et al. "Consciousness and the Double-Slit Interference Pattern: Six Experiments." *Physics Essays* 25, no. 2 (2012): 157–71.

Radin, Dean, et al. "Psychophysical Interactions With a Double-Slit Interference Pattern." *Physics Essays* 26, no. 4 (2013): 553–66.

Radin, Dean, et al. "Psychophysical Interactions With a Single-Photon Double-Slit Optical System." *Quantum Biosystems* 6, no. 1 (2015): 82–98.

Radin, Dean, et al. "Re-Examining Psychokinesis: Commentary on the Bösch, Steinkamp, and Boller Meta-Analysis." *Psychological Bulletin* 132 (2006): 529–32.

"Reality Is, As Tesla Said, 'Non-Physical': Why the Large Hadron Collider Will Never Find the 'God Particle.'" Collective Evolution website, November 15, 2016. http://www.collective-evolution.com/2016/11/15/most-of-reality-is-as-tesla-said-non-physical-why-the-large-hadron-collider-will-never-find-the-god-particle/.

"The Relativity of Space and Time." Einstein Online website, n.d. http://www.einstein-online.info/elementary/specialRT/relativity_space_time.html.

Ring, Kenneth, and Evelyn Elsaesser Valarino. *Lessons from the Light: What We Can Learn from the Near-Death Experience*. Needham, MA: Moment Point, 2006.

Ring, Kenneth, and Sharon Cooper. *Mindsight: Near-Death and Out-of-Body Experiences in the Blind*. New York: iUniverse, 2008.

Robinson, Andrew, and Philip Anderson. *Einstein: A Hundred Years of Relativity*. New York: Abrams, 2005.

Rosenblum, Bruce, and Fred Kuttner. *Quantum Enigma: Physics Encounters Consciousness*. Oxford: Oxford University Press, 2011.

Russell, Peter. *From Science to God: The Mystery of Consciousness and the Meaning of Light*. Novato, CA: New World Library, 2003.

Sagan, Carl. *The Demon-Haunted World: Science As a Candle in the Dark*. New York: Random House, 1995.

Sagan, Carl. *The Dragons of Eden: Speculations on the Evolution of Human Intelligence*. New York: Random House, 1977.

Saraswati, Swami Dayananda. *Value of Values*. Mylapore, India: Arsha Vidya Research and Publication Trust, 2007.

Schiller, F. C. S. *Riddles of the Sphinx: A Study in the Philosophy of Humanism*. Charleston, SC: Forgotten Books, 2010.

Schnabel, Jim. *Remote Viewers: The Secret History of America's Psychic Spies*. New York: Dell, 1997.

Schrödinger, Erwin. *My View of the World*. Woodbridge, CT: Ox Bow Press, 1983.

Schrödinger, Erwin. *What Is Life? With Mind and Matter and Autobiographical Sketches*. London: Cambridge University Press, 1969.

Schwartz, Gary. "Is Consciousness More than the Brain?" Gary Schwartz web-page, video, n.d. http://opensciences.org/gary-schwartz.

Schwartz, Gary. *The Sacred Promise: How Science Is Discovering Spirit's Collaboration with Us in Our Daily Lives*. New York: Atria, 2014.

Schwartz, Gary, and William Simon. *The Afterlife Experiments: Breakthrough Scientific Evidence of Life After Death*. New York: Atria, 2003.

Schwartz, Stephan. "The New Map in the Study of Consciousness." In *What Is Reality: The New Map of Cosmos and Consciousness* by Ervin Laszlo. New York: SelectBooks, 2016.

Schwartz, Stephan. *Opening to the Infinite: The Art and Science of Nonlocal Awareness*. Buda, TX: Nemoseen Media, 2007.

Schwartz, Stephan. "Through Time and Space: The Evidence for Remote Viewing." In *Evidence for Psi: Thirteen Empirical Research Reports*, edited by Damien Broderick and Ben Goertzel. Jefferson, NC: McFarland, 2015.

"The Science of Consciousness [meeting program]." Center for Consciousness Studies, University of Arizona, 2017. http://www.consciousness.arizona.edu/documents/TSC2017AbstractBook-final-5.10.17-final.pdf

"Scientists Found That The Soul Doesn't Die: It Goes Back To The Universe." Peace Quarters website, n.d. https://www.peacequarters.com/scientists-found-soul-doesnt-die-goes-back-universe/.

Scully, Suzannah. "Diane Hennacy Powell: Telepathy and the ESP Enigma." Cosmos in You website, May 9, 2017. https://www.suzannahscully.com/full-episodes.

Scully, Suzannah. "Donald Hoffman: Do We See Reality As It Really Is?." Cosmos in You website, July 27, 2015. https://www.suzannahscully.com/full-episodes.

"Selected Psi Research Publications." Dean Radin website, n.d. http://deanradin.com/evidence/evidence.htm.

Sheldrake, Rupert. *Dogs That Know When Their Owners Are Coming Home.* New York, New York: Crown, 1999.

Sheldrake, Rupert. *Science Set Free: Dispelling Dogma.* New York: Random House, 2012.

Sheldrake, Rupert. *The Sense of Being Stared At: And Other Aspects of the Extended Mind.* New York: Crown, 2003.

Sheldrake, Rupert. "The Sense of Being Stared At, Part I: Is it Real or Illusory?" *Journal of Consciousness Studies* 12, no. 6 (2005): 10–31.

Sheldrake, Rupert. "Sir John Maddox: Book for Burning." Rupert Sheldrake webpage, n.d. https://www.sheldrake.org/reactions/sir-john-maddox-book-for-burning.

Sheldrake, Rupert, and Pamela Smart. "Jaytee: A Dog Who Knew When His Owner Was Coming Home; The ORF Experiment." Rupert Sheldrake webpage, video, n.d. http://www.sheldrake.org/videos/jaytee-a-dog-who-knew-when-his-owner-was-coming-home-the-orf-experiment.

Sheldrake, Rupert, and Pamela Smart. "A Dog That Seems to Know When His Owner Is Coming Home: Videotaped Experiments and Observations." *Journal of Scientific Exploration* 14, (2000): 233–55.

Shermer, Michael. "Why Climate Skeptics Are Wrong." *Scientific American*, December 1, 2015. https://www.scientificamerican.com/article/why-climate-skeptics-are-wrong/.

Sidgwick, H. "Address by the President at the First General Meeting." *Proceedings of the Society for Psychical Research* (1882).

Smolin, Lee. *Time Reborn: From the Crisis in Physics to the Future of the Universe*. Boston: Houghton Mifflin Harcourt, 2013.

Sommer, A. "Psychical Research in the History and Philosophy of Science: An Introduction and Review." *Studies in History and Philosophy of Biological and Biomedical Sciences* 48 (2014): 38–45. https://www.sciencedirect.com/science/article/pii/S1369848614001204#!.

Spira, Rupert. *The Nature of Consciousness: Essays on the Unity of Mind and Matter*. Oakland, CA: New Harbinger, 2017.

Stapp, Henry. *Quantum Theory and Free Will: How Mental Intentions Translate into Bodily Actions*. New York: Springer, 2017.

"Stargate Project." Wikipedia, last updated August 19, 2017. https://en.wikipedia.org/wiki/Stargate_Project.

Statt, Nick. "Elon Musk Launches Neuralink, A Venture To Merge the Human Brain With AI." *The Verge*, March 27, 2017. https://www.theverge.com/2017/3/27/15077864/elon-musk-neuralink-brain-computer-interface-ai-cyborgs.

Stevenson, Ian. *Children Who Remember Previous Lives: A Question of Reincarnation*. Jefferson, NC: McFarland, 2001.

Stevenson, Ian. *Reincarnation and Biology: A Contribution to the Etiology of Birthmarks and Birth Defects*. Vol. 1, *Birthmarks*. Westport, CT: Praeger, 1997.

Stevenson, Ian. *Where Reincarnation and Biology Intersect*. Westport, CT: Praeger, 1997.

Stott, Shane. *The Float Tank Cure: Free Yourself from Stress, Anxiety, and Pain the Natural Way*. Tecumseh, MI: DiggyPOD, 2015.

Swanson, Claude. *The Synchronized Universe: New Science of the Paranormal*. Vol. 1. Tuscon, AZ: Poseidia, 2003.

Swanson, Claude. *The Synchronized Universe: New Science of the Paranormal*. Vol. 2, *Life Force, the Scientific Basis: Breakthrough Physics of Energy Medicine, Healing, Chi, and Quantum Consciousness*. Tuscon, AZ: Poseidia, 2009.

Talbot, Michael. *The Holographic Universe: The Revolutionary Theory of Reality*. New York: HarperCollins, 1991.

Targ, Russell. *The Reality of ESP: A Physicist's Proof of Psychic Phenomena*. Wheaton, IL: Quest, 2012.

Targ, Russell, and Keith Harary. *The Mind Race: Understanding and Using Psychic Abilities*. New York: Villard, 1984.

"Telepathy." Wikipedia, last updated July 16, 2017. https://en.wikipedia.org/wiki/Telepathy.

Through the Wormhole. "Is There Life After Death?" Season 2, episode 1. Science Channel, June 8, 2011.

Tiller, William. *Psychoenergetic Science: A Second Copernican-Scale Revolution*. Walnut Creek, CA: Pavior, 2007.

Tiller, William, and Walter Dibble. "A Brief Introduction to Intention-Host Device Research." Unpublished paper, William A. Tiller Foundation. https://www.tillerinstitute.com/pdf/White%20Paper%20I.pdf.

Treffert, Darold. *Islands of Genius: The Bountiful Mind of the Autistic, Acquired, and Sudden Savant*. London: Jessica Kingsley, 2012.

Tressoldi, P. E. "Extraordinary Claims Require Extraordinary Evidence: The Case of Non-Local Perception, a Classical and Bayesian Review of Evidences." *Frontiers in Psychology* 2, no. 117 (2011).

Tsakiris, Alex. *Why Science Is Wrong...About Almost Everything*. San Antonio, TX: Anomalist, 2014.

Tucker, Jim. *Return to Life: Extraordinary Cases of Children Who Remember Past Lives*. New York: St. Martin's, 2013.

Turing, Alan Mathison. "Computing Machinery and Intelligence." *Mind: A Quarterly Review of Psychology and Philosophy* LIX, no. 236 (1950): 433–60.

"2007 Distance Healing Experiment with Pennsylvania State University Medical School." Natural Healing Center website, n.d. http://www.naturalhealingcenter.com/creative/jixingli.htm.

Tyson, Neil deGrasse. Twitter post. June 14, 2013. https://twitter.com/neiltyson/status/345551599382446081?lang=en.

Urban, Tim. "Neuralink and the Brain's Magical Future." Wait But Why website, April 20, 2017. https://waitbutwhy.com/2017/04/neuralink.html.

Utts, Jessica. "Appreciating Statistics." *Journal of the American Statistical Association* 111 (2016): 1373–80. https://www.tandfonline.com/doi/full/10.1080/01621459.2016.1250592.

Utts, Jessica. "An Assessment of the Evidence for Psychic Functioning." *Journal of Parapsychology* 59, no. 4 (1995): 289–320.

Utts, Jessica. "An Assessment of the Evidence for Psychic Functioning." Division of Statistics, University of California, Davis, 1995. http://www.ics.uci.edu/~jutts/air.pdf

Utts, Jessica. "From Psychic Claims to Science: Testing Psychic Phenomena with Statistics." Department of Statistics, University of California, Davis, August 3, 2006. http://www.ics.uci.edu/~jutts/Sweden.pdf.

Utts, Jessica. "Home Page for Professor Jessica Utts." Department of Statistics, University of California, Irvine. http://www.ics.uci.edu/~jutts/.

van Lommel, Pim. *Consciousness Beyond Life: The Science of the Near-Death Experience*. New York: HarperOne, 2010.

van Lommel, Pim. "Near-Death Experience, Consciousness, and the Brain: A New Concept about the Continuity of Our Consciousness Based on Recent Scientific Research on Near-Death Experience in Survivors of Cardiac Arrest." *World Futures* 62 (2006): 134–51. http://deanradin.com/evidence/vanLommel2006.pdf.

van Lommel, Pim, et al. "Near-Death Experience in Survivors of Cardiac Arrest: A Prospective Study in the Netherlands." *The Lancet* 9298 (2001): 2039–45. http://www.thelancet.com/journals/lancet/article/PIIS0140673601071008/fulltext.

Warman, Matt. "Stephen Hawking Tells Google 'Philosophy Is Dead.'" *The Telegraph*, May 17, 2011. http://www.telegraph.co.uk/technology/google/8520033/Stephen-Hawking-tells-Google-philosophy-is-dead.html.

Watt, C. "Precognitive Dreaming: Investigating Anomalous Cognition and Psychological Factors." *Journal of Parapsychology* 78, no. 1 (2014): 115–25.

Weinberg, Steven. *Dreams of a Final Theory*. London: Vintage, 2010.

Weiss, Brian. *Many Lives, Many Masters: The True Story of a Prominent Psychiatrist, His Young Patient, and the Past-Life Therapy That Changed Both Their Lives*. New York: Simon & Schuster, 2012.

Weiss, Brian. *Same Soul, Many Bodies: Discover the Healing Power of Future Lives Through Progression Therapy*. New York: Free Press, 2004.

Wheeler, John Archibald. "Law without Law." In *Quantum Theory and Measurement* edited by John Archibald Wheeler and Wojciech Hubert Zurek. Princeton, NJ: Princeton University Press, 1983.

Wheeler, John Archibald, and Wojciech Hubert Zurek, eds. *Quantum Theory and Measurement*. Princeton, NJ: Princeton University Press, 1984.

Wigner, Eugene Paul. "The Place of Consciousness in Modern Physics." In *Philosophical Reflections and Syntheses*, edited by Eugene Paul Wigner and Jagdish Mehra. Berlin: Springer, 1997.

Wigner, Eugene Paul. "Remarks on the Mind-Body Problem." In *The Scientist Speculates*, edited by I. J. Good, pp. 284–302. London: Heinemann, 1961.

Wigner, Eugene. *Symmetries and Reflections*. Woodbridge, CT: Ox Bow Press, 1979.

"Wikipedia Under Threat." Rupert Sheldrake webpage, n.d. https://www.sheldrake.org/about-rupert-sheldrake/blog/wikipedia-under-threat.

William A. Tiller Institute for Psychoenergetic Science. http://www.tillerinstitute.com/.

Williams, B. J. "Revisiting the Ganzfeld ESP Debate: A Basic Review and Assessment." *Journal of Scientific Exploration* 25, no. 4 (2011): 639–61.

About the Author

Mark Gober is an author whose worldview was turned upside down in late 2016 when he was exposed to world-changing science. After researching extensively, he wrote *An End to Upside Down Thinking* to introduce the general public to these cutting-edge ideas—all in an effort to encourage a much-needed global shift in scientific and existential thinking. Mark is a senior member of Sherpa Technology Group in Silicon Valley, a firm that advises technology companies on mergers and acquisitions and business strategy. He previously worked as an investment banking analyst in New York. Mark has been quoted for his opinions on business and technology matters in *Bloomberg Businessweek* and elsewhere, and he has authored internationally published business articles. He graduated *magna cum laude* from Princeton University, where he was captain of the tennis team.

www.markgober.com